vectors

vectors

raymond a. barnett

instructor of mathematics

oakland city college

john n. fujii

instructor of mathematics

oakland city college

john wiley & sons, inc., new york · london

Printed in the United States of America

preface

This supplement should be useful to lower-division students in mathematics, physics, and engineering and, in addition, to mathematically capable students in high school. We have found, in teaching the calculus, that students have a need for a supplementary source of material on the algebra of vectors devoid of other complicating topics. We also have reason to believe that many engineering and physics instructors feel the same way with regard to certain courses in their field.

The elementary notions of analytic geometry and the standard high school courses in trigonometry and intermediate algebra are all that are necessary for profitable study of the material presented here. No calculus is required, for the development stops at the threshold of the vector calculus. The last chapter, however, includes examples and exercises of a transitional nature which are designed to prepare the reader for a natural entry into the vector calculus.

The topics included in this supplement are neither new nor different but are, we hope, presented with continuity and clarity. Numerous examples and exercises are included so that readers will have an opportunity to make the development their own. Additional features are (a) definitions and theorems are "set off" and numbered for easy reference; (b) a brief summary of

v

the important definitions and theorems is included before each set of problems to provide the reader with an overview of the preceding topics; (c) the problems and exercises are "paired," where possible, into even and odd, with answers given to all items. In addition, the problems are grouped into general, geometric, and physics-engineering types.

RAYMOND A. BARNETT
JOHN. N. FUJII

Oakland, California
November, 1962

contents

1

introduction

1.1 VECTORS AND SCALARS

Mathematicians and physicists deal with many kinds of quantities. Some quantities, such as mass, volume, distance, and temperature, can be completely characterized by a single real number. Other quantities, such as force, velocity, displacement, and acceleration, require a direction and a magnitude for their specification. The following definitions will distinguish between these quantities.

DEFINITION 1.1.1 *A* **scalar** *is a quantity that can be completely characterized by a single real number.*

DEFINITION 1.1.2 *A* (**free**) **vector** *is a quantity that requires for its complete specification a magnitude and a direction.*

1

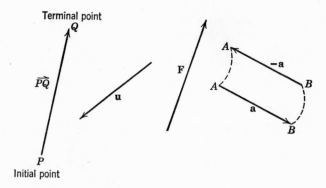

FIGURE 1.1.1   Vector notation.

A vector can be represented geometrically by a directed line segment in space, that is, if P and Q are distinct points, then the directed line segment from P to Q is called a vector.   The point P is called the **initial point** or end and Q the **terminal point** or end of the vector.   If the initial and terminal ends of a vector are important in a discussion, a symbol such as "$\overrightarrow{PQ}$" will be used to denote the vector with initial point P and terminal point Q; otherwise a single boldface letter, such as **a**, **u**, **F**, will be used to denote vectors.   If **a** is used to denote the vector with initial point A and terminal point B (i.e., $\overrightarrow{AB}$), then $-$**a** will denote the vector with initial point B and terminal point A (i.e., $\overrightarrow{BA}$).

The **magnitude** of a vector $\overrightarrow{AB}$ is a positive scalar quantity that is represented by the length of the directed line segment joining points A and B.   "$|$**a**$|$" or "a" will be used to denote the magnitude of the vector **a** and "$|AB|$" will be used to denote the magnitude of the vector $\overrightarrow{AB}$.   (*Note:* $|$**a**$| \geq 0$ for all vectors **a**.)

DEFINITION 1.1.3   **Equality of (free) vectors.**

$$[\mathbf{a} = \mathbf{b}] \Leftrightarrow \left[\begin{array}{l} \text{(i)} \; \textit{direction of } \mathbf{a} = \textit{direction of } \mathbf{b.} \\ \text{(ii)} \; \textit{magnitude of } \mathbf{a} = \textit{magnitude of } \mathbf{b.} \end{array} \right]^{*}$$

* The symbol "$\Leftrightarrow$" used in Definition 1.1.3 means "if and only if."   It will be used wherever appropriate for brevity and convenience.

The above definition of equality implies that a vector **a** can be represented by infinitely many directed line segments in space and hence has no fixed position in space. This is the reason for the word "free" in the definition. For certain applications in physics and geometry it is useful to introduce the concepts of **sliding** and **bound** vectors. A sliding vector is restricted to a line of action; hence two sliding vectors are said to be equal if and only if they have the same magnitude and direction and lie on the same line. A bound or fixed vector is restricted to a special point of application. Two bound vectors are said to be equal if and only if they have the same magnitude, direction, and the same initial point.

In the development that is to follow, vectors will be interpreted as fixed, sliding, or free as is appropriate. Unless otherwise

TABLE 1.1.1
Comparison of Statements in Vector and Nonvector Form

Vector Form	Nonvector Form	Description
$\overrightarrow{AB} = (\overrightarrow{AP})t$	$x = x_1 + (x_2 - x_1)t$ $y = y_1 + (y_2 - y_1)t$ $z = z_1 + (z_2 - z_1)t$	The line through $A(x_1, y_1, z_1)$ and $B(x_2, y_2, z_2)$
$\mathbf{n} \cdot \overrightarrow{OP} = 0$	$n_x x + n_y y + n_z z = k$	A plane perpendicular to **n**
$|\overrightarrow{OP}| = k$	$(x - x_0)^2 + (y - y_0)^2 + (z - z_0)^2 = k^2$	A sphere with center O and radius k
$\mathbf{a} \cdot \mathbf{x} = \alpha$ $\mathbf{b} \cdot \mathbf{x} = \beta$ $\mathbf{c} \cdot \mathbf{x} = \gamma$	$a_1 x_1 + a_2 x_2 + a_3 x_3 = \alpha$ $b_1 x_1 + b_2 x_2 + b_3 x_3 = \beta$ $c_1 x_1 + c_2 x_2 + c_3 x_3 = \gamma$	A system of linear equations
$\Sigma \mathbf{F}_i = \mathbf{0}$ $\Sigma \mathbf{M}_i = \mathbf{0}$	$\Sigma X_n = \Sigma Y_n = \Sigma Z_n = 0$ and $\Sigma(y_n Z_n - z_n Y_n)$ $= \Sigma(z_n X_n - x_n Z_n)$ $= \Sigma(x_n Y_n - y_n X_n) = 0$	The conditions for static equilibrium

stated, however, all vectors will be free vectors, and the word "free" will henceforth be omitted from general use.

DEFINITION 1.1.4   *The* **zero** *or* **null vector**, *denoted by* **0**, *will be assigned zero magnitude and arbitrary direction.*

The reader may wonder about the apparent carelessness in defining a direction for the zero vector. As will be seen later, the above definition will eliminate the need for stating exceptional cases when writing certain relationships in vector form.

Vectors with properties yet to be discussed are of considerable use. Although an appreciation of their value will only come after some experience, we note two benefits derived from working within a vector framework.

a. Vectors provide a concise and easy-to-follow symbolism for the expression of useful physical relationships and formulas. (See Table 1.1.1.)

b. Vectors enable one to analyze and solve certain geometric and physical problems without special reference to a particular coordinate system.

1.2   VECTOR ADDITION AND MULTIPLICATION OF A VECTOR BY A SCALAR

The definition of the sum of two vectors given below is equivalent to the parallelogram law used in physics for combining forces and velocities.

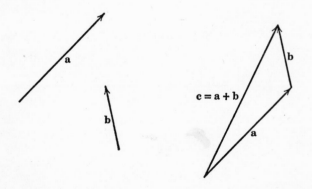

FIGURE 1.2.1   Addition of vectors.

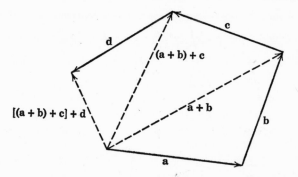

FIGURE 1.2.2   Addition of several vectors.

DEFINITION 1.2.1   **Vector Addition.**   *Given two vectors* **a** *and* **b**, *the vector* **c**, *called* **the sum (or resultant)** *of* **a** *and* **b**, *is obtained by placing the initial end of* **b** *on the terminal end of* **a** *and constructing the vector* **c** *so that its initial end will be at the initial end of* **a** *and its terminal end will be at the terminal end of* **b**.

The **sum (or resultant) of several vectors** can be found by repeated application of Definition 1.2.1.

DEFINITION 1.2.2   **Vector Subtraction.**   *The difference of two vectors* **a** *and* **b**, *denoted by* **a** − **b**, *is a vector* **c** *such that* **a** = **b** + **c**, *that is,*

$$[\mathbf{a} - \mathbf{b} = \mathbf{c}] \Leftrightarrow [\mathbf{a} = \mathbf{b} + \mathbf{c}].$$

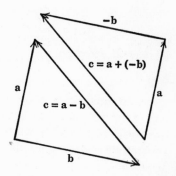

FIGURE 1.2.3   Vector subtraction.

Definition 1.2.2 may be more readily understood by recalling the analogous definition for the difference of two real numbers. The difference of two vectors is also defined in terms of addition by the use of the symbol "−**a**," that is, **a** − **b** = **a** + (−**b**). Hence subtracting a vector is the same as adding its negative. Both forms of the definition are useful in applications.

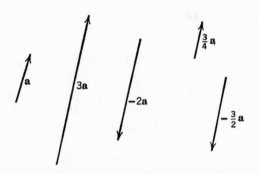

FIGURE 1.2.4   Scalar multiples.

There are several different types of multiplication involving vectors; each has a distinct geometric interpretation. The first type of product considered is that of a scalar and a vector. (Two others types of products will be introduced in Section 2.1 and Section 3.1.)

DEFINITION 1.2.3 **Multiplication of a vector by a scalar.** *If h is a scalar and* **a** *is a vector, then* **ha** *is a vector with magnitude* $|h|\,|a|$ *and direction* $\begin{cases} \text{the same as } \mathbf{a} \text{ if } h > 0 \\ \text{the same as } -\mathbf{a} \text{ if } h < 0 \end{cases}$ *and arbitrary if* $h = 0$.

Vector quantities obey certain algebraic laws. These laws enable us to manipulate the symbols which represent vector quantities in much the same way as the symbols that represent real numbers are manipulated in the algebra of real numbers. We summarize the important laws which arise as a consequence of the definitions stated above.

ALGEBRAIC LAWS (**a**, **b**, **c** are vectors; m and n are scalars)

1.2.1	$\mathbf{a} + \mathbf{b} = \mathbf{b} + \mathbf{a}$	commutative law
1.2.2	$\mathbf{a} + (\mathbf{b} + \mathbf{c}) = (\mathbf{a} + \mathbf{b}) + \mathbf{c}$	associative law
1.2.3	$\mathbf{a} + \mathbf{0} = \mathbf{a}$	additive identity
1.2.4	$\mathbf{a} + (-\mathbf{a}) = \mathbf{0}$	additive inverse
1.2.5	$1\mathbf{a} = \mathbf{a}$	unit element
1.2.6	$0\mathbf{a} = \mathbf{0}$	zero element
1.2.7	$(mn)\mathbf{a} = m(n\mathbf{a})$	associative law

1.2.8 $(m + n)\mathbf{a} = m\mathbf{a} + n\mathbf{a}$ distributive law

1.2.9 $m(\mathbf{a} + \mathbf{b}) = m\mathbf{a} + m\mathbf{b}$ distributive law

Note that no distributive law involving vector quantities only (i.e., $\mathbf{a}(\mathbf{b} + \mathbf{c}) = \mathbf{ab} + \mathbf{ac}$) is included in the preceding list. The reason is that a product of two vectors has not as yet been defined. The proof of Laws 1.2.1 and 1.2.9 will be left as exercises. Law 1.2.2 can be established by a construction. The remaining laws follow directly from the given definitions.

The magnitudes (i.e., the absolute values) of vectors also have certain properties which are summarized below.

ALGEBRAIC LAWS for the magnitude of vectors ($\mathbf{a}$ and $\mathbf{b}$ are vectors, m a scalar)

1.2.10 $|\mathbf{a}| \geq 0, [|\mathbf{a}| = 0] \Leftrightarrow [\mathbf{a} = \mathbf{0}]$

1.2.11 $|\mathbf{a}| + |\mathbf{b}| \geq |\mathbf{a} + \mathbf{b}|$ triangle inequality

1.2.12 $|\mathbf{a}| - |\mathbf{b}| \leq |\mathbf{a} - \mathbf{b}|$ triangle inequality

1.2.13 $|m\mathbf{a}| = |m|\,|\mathbf{a}|$

The following examples illustrate how the laws and properties just stated are utilized. The reader is encouraged to read each example carefully, identifying the appropriate law or definition that justifies each step of the procedure.

EXAMPLE 1.2.1   If M is the midpoint of the segment AB and O is any point in space, show that $\overrightarrow{OM} = \frac{1}{2}\overrightarrow{OA} + \frac{1}{2}\overrightarrow{OB}$.

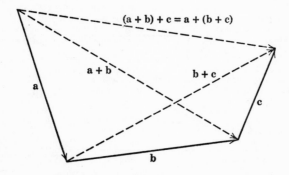

FIGURE 1.2.5   The associative law.

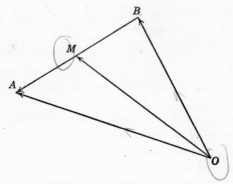

FIGURE 1.2.6   Midpoint of AB.

Solution

a. $\vec{BA} = \vec{OA} - \vec{OB}$ Definition 1.2.2

b. $\vec{OM} = \vec{OB} + \frac{1}{2}\vec{BA}$ Definition 1.2.1

c. $\vec{OM} = \vec{OB} + \frac{1}{2}(\vec{OA} - \vec{OB})$ a., b.

d. $\vec{OM} = \vec{OB} + (\frac{1}{2}\vec{OA} - \frac{1}{2}\vec{OB})$ Law 1.2.9

e. $\vec{OM} = (\vec{OB} - \frac{1}{2}\vec{OB}) + \frac{1}{2}\vec{OA}$ Laws 1.2.1, 1.2.2

f. $\vec{OM} = (1 - \frac{1}{2})\vec{OB} + \frac{1}{2}\vec{OA}$ Laws 1.2.5, 1.2.8

g. Therefore $\vec{OM} = \frac{1}{2}\vec{OA} + \frac{1}{2}\vec{OB}$ Law 1.2.1

EXAMPLE 1.2.2   Show that the diagonals of a parallelogram bisect each other.

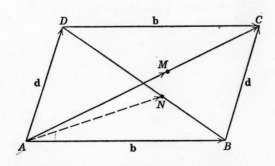

FIGURE 1.2.7   Diagonals of a parallelogram.

Discussion

Let the parallelogram have vertices A, B, C, D, as shown in Figure 1.2.7. Let M and N be the midpoints of AC and BD, respectively. We must show that M and N coincide. Two methods are presented here, and it is suggested that the reader try both methods in one or two other problems.

Solution (method I, Figure 1.2.7)

a. $\overrightarrow{AN} = \frac{1}{2}\mathbf{d} + \frac{1}{2}\mathbf{b}$ Example 1.2.1

b. $\overrightarrow{AM} = \frac{1}{2}\overrightarrow{AC}$ Definition 1.2.3

c. $\overrightarrow{AC} = \mathbf{b} + \mathbf{d}$ Definition 1.2.1

d. $\overrightarrow{AM} = \frac{1}{2}(\mathbf{b} + \mathbf{d})$ b., c.

e. $\overrightarrow{AM} = \frac{1}{2}\mathbf{b} + \frac{1}{2}\mathbf{d}$ Law 1.2.9

f. $\overrightarrow{AM} = \overrightarrow{AN}$ a., e., Law 1.2.1

g. Therefore since $\overrightarrow{AM}$ and $\overrightarrow{AN}$ have the same initial point, M and N must coincide.   Definition 1.1.3

Solution (method II, Figure 1.2.8)

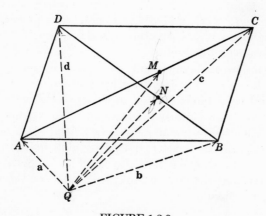

FIGURE 1.2.8

a. Let Q be an arbitrary point, $\mathbf{a} = \overrightarrow{QA}$, $\mathbf{b} = \overrightarrow{QB}$, $\mathbf{c} = \overrightarrow{QC}$, and $\mathbf{d} = \overrightarrow{QD}$.   Definition 1.1.2

b. $\overrightarrow{QN} = \frac{1}{2}\mathbf{d} + \frac{1}{2}\mathbf{b}$ Example 1.2.1

c. $\overrightarrow{QN} = \frac{1}{2}(\mathbf{d} + \mathbf{b})$ Law 1.2.9

d. $\overrightarrow{AB} = \overrightarrow{DC}$ Given

e. $\overrightarrow{AB} = \mathbf{b} - \mathbf{a}$ Definition 1.2.2

f. $\overrightarrow{DC} = \mathbf{c} - \mathbf{d}$ Definition 1.2.2

g. $\mathbf{b} - \mathbf{a} = \mathbf{c} - \mathbf{d}$ d., e., f.

h. $\mathbf{b} + \mathbf{d} = \mathbf{a} + \mathbf{c}$ Add $\mathbf{a} + \mathbf{d}$ to both
                                                     sides in step g,
                                                     Laws 1.2.1, 1.2.4

i. $\overrightarrow{QN} = \frac{1}{2}(\mathbf{a} + \mathbf{c})$ c., h.

j. $\overrightarrow{QN} = \frac{1}{2}\mathbf{a} + \frac{1}{2}\mathbf{c}$ Law 1.2.9

k. $\overrightarrow{QM} = \frac{1}{2}\mathbf{a} + \frac{1}{2}\mathbf{c}$ Example 1.2.1

l. $\overrightarrow{QN} = \overrightarrow{QM}$ j., k.

m. Therefore since $\overrightarrow{QM}$ and $\overrightarrow{QN}$ have the same initial point, M and N must coincide.   Definition 1.1.3

The reader should note how vectors are introduced into the solutions of problems in two distinct ways. In method I of Example 1.2.2, the line segments forming the figure were used directly to introduce vectors. The point A was arbitrarily chosen as an initial point, and appropriate vectors were defined accordingly. In method II of the example, an arbitrary initial point Q was chosen and connected to the critical points of the figure to form suitable vectors.

In practice one does not usually proceed as formally as above. However, it would be worthwhile for the reader to follow the more formal procedures in a few exercises so that the algebraic laws may be more clearly understood. No attempt should be made to memorize the above proofs. After some experience has been gained, the reader should be able to construct his own proofs. In carrying out the proof of a given theorem it is unlikely that the sequence of steps would be the same for two different individuals.

1.3   COLLINEAR AND COPLANAR VECTORS

DEFINITION 1.3.1   *Two vectors are said to be* **collinear** *if they are parallel to the same line.*

DEFINITION 1.3.2   *Three vectors are said to be* **coplanar** *if they are parallel to the same plane.*

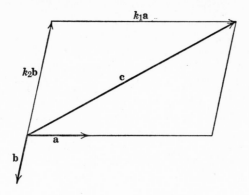

FIGURE 1.3.1 $\mathbf{c} = k_1\mathbf{a} + k_2\mathbf{b}$.

THEOREM 1.3.1 If vectors $\mathbf{a}$ and $\mathbf{b}$ are noncollinear, then any vector $\mathbf{c}$ coplanar with $\mathbf{a}$ and $\mathbf{b}$ can be represented as a linear combination of $\mathbf{a}$ and $\mathbf{b}$. That is, there are scalars k_1 and k_2 such that

$$\mathbf{c} = k_1\mathbf{a} + k_2\mathbf{b}.$$

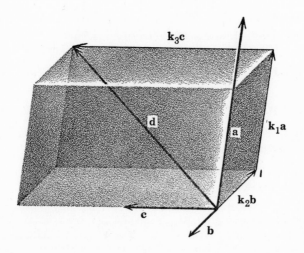

FIGURE 1.3.2 $\mathbf{d} = k_1\mathbf{a} + k_2\mathbf{b} + k_3\mathbf{c}$.

Discussion

Consider representative vectors **a**, **b**, and **c** with a common initial point. Construct a parallelogram with diagonal **c** and sides parallel to **a** and **b**. $k_1\mathbf{a}$ and $k_2\mathbf{b}$ are represented by the sides of the parallelogram.

THEOREM 1.3.2 If vectors **a**, **b**, and **c** are noncoplanar, every vector **d** in 3-space can be represented as a linear combination of **a**, **b**, and **c**, that is, there are scalars k_1, k_2, and k_3 such that

$$\mathbf{d} = k_1\mathbf{a} + k_2\mathbf{b} + k_3\mathbf{c}.$$

Discussion

Consider representative vectors **a**, **b**, **c**, and **d** with a common initial point. Construct a parallelepiped with diagonal **d** and edges parallel to **a**, **b**, and **c**. $k_1\mathbf{a}$, $k_2\mathbf{b}$, and $k_3\mathbf{c}$ are represented by the edges of the parallelepiped.

DEFINITION 1.3.3 *A set of n vectors* $\mathbf{v}_1$, $\mathbf{v}_2$, . . . , $\mathbf{v}_n$ *is said to be* **linearly dependent** *if there are scalars* k_1, k_2, . . . , k_n, *not all zero, such that* $k_1\mathbf{v}_1 + k_2\mathbf{v}_2 + \cdots + k_n\mathbf{v}_n = \mathbf{0}$; *otherwise the set of vectors is said to be* **linearly independent.***

THEOREM 1.3.3 $\begin{bmatrix} \mathbf{v}_1 \text{ and } \mathbf{v}_2 \text{ are} \\ \text{linearly dependent} \end{bmatrix} \Leftrightarrow \begin{bmatrix} \mathbf{v}_1 \text{ and } \mathbf{v}_2 \text{ are} \\ \text{collinear} \end{bmatrix}$

THEOREM 1.3.4 $\begin{bmatrix} \mathbf{v}_1 \text{ and } \mathbf{v}_2 \text{ are} \\ \text{linearly independent} \end{bmatrix} \Leftrightarrow \begin{bmatrix} \mathbf{v}_1 \text{ and } \mathbf{v}_2 \text{ are} \\ \text{noncollinear} \end{bmatrix}$

THEOREM 1.3.5 $\begin{bmatrix} \mathbf{v}_1, \mathbf{v}_2, \text{ and } \mathbf{v}_3 \text{ are} \\ \text{linearly dependent} \end{bmatrix} \Leftrightarrow \begin{bmatrix} \mathbf{v}_1, \mathbf{v}_2, \text{ and } \mathbf{v}_3 \text{ are} \\ \text{coplanar} \end{bmatrix}$

THEOREM 1.3.6 $\begin{bmatrix} \mathbf{v}_1, \mathbf{v}_2, \text{ and } \mathbf{v}_3 \text{ are} \\ \text{linearly independent} \end{bmatrix} \Leftrightarrow \begin{bmatrix} \mathbf{v}_1, \mathbf{v}_2, \text{ and } \mathbf{v}_3 \text{ are} \\ \text{noncoplanar} \end{bmatrix}$

Question: Which of the above theorems are equivalent?

THEOREM 1.3.7 Three or more vectors in 2-dimensional space are linearly dependent.

THEOREM 1.3.8 Four or more vectors in 3-dimensional space are linearly dependent.

* In a more general development of vectors the notion of linear dependence and independence is introduced and utilized to a considerable extent in the further development of the subject. The definition of linear dependence and its relationship to collinear and coplanar vectors are stated here mainly for the benefit of those who will pursue the subject further.

SUMMARY

1.1 Vectors and Scalars

DEFINITION 1.1.1 *A* **scalar** *is a quantity that can be completely characterized by a single real number.*

DEFINITION 1.1.2 *A* **vector** *is a quantity that requires for its complete specification a magnitude and a direction.*

DEFINITION 1.1.3 **Equality of vectors.**

$$[\mathbf{a} = \mathbf{b}] \Leftrightarrow \left[\begin{array}{l} \text{(i)} \ \ direction \ of \ \mathbf{a} = direction \ of \ \mathbf{b} \\ \text{(ii)} \ \ magnitude \ of \ \mathbf{a} = magnitude \ of \ \mathbf{b} \end{array} \right]$$

DEFINITION 1.1.4 *The* **zero or null vector,** *denoted by* **0,** *will be assigned zero magnitude and arbitrary direction.*

1.2 Vector Addition and Multiplication of a Vector by a Scalar

DEFINITION 1.2.1 **Vector Addition.** *Given two vectors* **a** *and* **b** *the vector* **c,** *called the sum of* **a** *and* **b,** *is obtained by placing the initial end of* **b** *on the terminal end of* **a** *and constructing the vector* **c** *so that its initial end will be at the initial end of* **a** *and its terminal end will be at the terminal end of* **b.**

DEFINITION 1.2.2 **Vector Subtraction.** *The difference of two vectors* **a** *and* **b,** *denoted by* **a** − **b,** *is a vector* **c** *such that* **a** = **b** + **c.** *Alternatively, we can write: Subtracting a vector is the same as adding its negative, that is,* **a** − **b** = **a** + (−**b**).

DEFINITION 1.2.3 **Multiplication of a vector by a scalar.** *If* h *is a scalar and* **a** *is a vector, then* h**a** *is a vector with magnitude* $|h| \, |\mathbf{a}|$ *and direction* $\begin{cases} the \ same \ as \ \mathbf{a} \ if \ h > 0 \\ the \ same \ as \ -\mathbf{a} \ if \ h < 0 \end{cases}$ *and arbitrary if* $h = 0$.

ALGEBRAIC LAWS (**a**, **b**, **c** are vectors; m and n are scalars)

1.2.1	$\mathbf{a} + \mathbf{b} = \mathbf{b} + \mathbf{a}$	commutative law
1.2.2	$\mathbf{a} + (\mathbf{b} + \mathbf{c}) = (\mathbf{a} + \mathbf{b}) + \mathbf{c}$	associative law
1.2.3	$\mathbf{a} + \mathbf{0} = \mathbf{a}$	additive identity
1.2.4	$\mathbf{a} + (-\mathbf{a}) = \mathbf{0}$	additive inverse
1.2.5	$1\mathbf{a} = \mathbf{a}$	unit element
1.2.6	$0\mathbf{a} = \mathbf{0}$	zero element
1.2.7	$(mn)\mathbf{a} = m(n\mathbf{a})$	associative law

1.2.8 $(m + n)\mathbf{a} = m\mathbf{a} + n\mathbf{a}$ distributive law
1.2.9 $m(\mathbf{a} + \mathbf{b}) = m\mathbf{a} + m\mathbf{b}$ distributive law
1.2.10 $|\mathbf{a}| \geq 0, [|\mathbf{a}| = 0] \Leftrightarrow [\mathbf{a} = \mathbf{0}]$
1.2.11 $|\mathbf{a}| + |\mathbf{b}| \geq |\mathbf{a} + \mathbf{b}|$ triangle inequality
1.2.12 $|\mathbf{a}| - |\mathbf{b}| \leq |\mathbf{a} - \mathbf{b}|$ triangle inequality
1.2.13 $|m\mathbf{a}| = |m|\,|\mathbf{a}|$

1.3   Collinear and Coplanar Vectors

DEFINITION 1.3.1   *Two vectors are said to be* **collinear** *if they are parallel to the same line.*

DEFINITION 1.3.2   *Three vectors are said to be* **coplanar** *if they are parallel to the same plane.*

DEFINITION 1.3.3   *A set of n vectors* $\mathbf{v}_1, \mathbf{v}_2, \ldots, \mathbf{v}_n$ *is said to be* **linearly dependent** *if there are scalars* $k_1, k_2, \ldots, k_n$, *not all zero, such that* $k_1\mathbf{v}_1 + k_2\mathbf{v}_2 + \cdots + k_n\mathbf{v}_n = \mathbf{0}$; *otherwise the set of vectors is said to be* **linearly independent.** ($n \geqslant 4$)

THEOREM 1.3.1   If vectors **a** and **b** are noncollinear, any vector **c** coplanar with **a** and **b** can be represented as a linear combination of **a** and **b**, that is, there are scalars k_1 and k_2 such that

$$\mathbf{c} = k_1\mathbf{a} + k_2\mathbf{b}.$$

THEOREM 1.3.2   If vectors **a**, **b**, and **c** are noncoplanar, every vector **d** in 3-space can be represented as a linear combination of **a**, **b**, and **c**, that is, there are scalars k_1, k_2, and k_3 such that

$$\mathbf{d} = k_1\mathbf{a} + k_2\mathbf{b} + k_3\mathbf{c}.$$

THEOREM 1.3.3 $\begin{bmatrix} \mathbf{v}_1 \text{ and } \mathbf{v}_2 \text{ are} \\ \text{linearly dependent} \end{bmatrix} \Leftrightarrow \begin{bmatrix} \mathbf{v}_1 \text{ and } \mathbf{v}_2 \text{ are} \\ \text{collinear} \end{bmatrix}$

THEOREM 1.3.4 $\begin{bmatrix} \mathbf{v}_1 \text{ and } \mathbf{v}_2 \text{ are} \\ \text{linearly independent} \end{bmatrix} \Leftrightarrow \begin{bmatrix} \mathbf{v}_1 \text{ and } \mathbf{v}_2 \text{ are} \\ \text{noncollinear} \end{bmatrix}$

THEOREM 1.3.5 $\begin{bmatrix} \mathbf{v}_1, \mathbf{v}_2, \text{ and } \mathbf{v}_3 \text{ are} \\ \text{linearly dependent} \end{bmatrix} \Leftrightarrow \begin{bmatrix} \mathbf{v}_1, \mathbf{v}_2, \text{ and } \mathbf{v}_3 \\ \text{are coplanar} \end{bmatrix}$

THEOREM 1.3.6 $\begin{bmatrix} \mathbf{v}_1, \mathbf{v}_2, \text{ and } \mathbf{v}_3 \text{ are} \\ \text{linearly independent} \end{bmatrix} \Leftrightarrow \begin{bmatrix} \mathbf{v}_1, \mathbf{v}_2, \text{ and } \mathbf{v}_3 \\ \text{are noncoplanar} \end{bmatrix}$

THEOREM 1.3.7   Three or more vectors in 2-dimensional space are linearly dependent.

THEOREM 1.3.8   Four or more vectors in 3-dimensional space are linearly dependent.

PROBLEM SET #1

A. General

1. Copy the coplanar vectors given in Figure #1-1 and construct the vectors

   (a) $\mathbf{a} + \mathbf{b}$ (b) $\mathbf{a} - \mathbf{c}$ (c) $(\mathbf{a} + \mathbf{b}) - (\mathbf{a} - \mathbf{c})$ (d) $2\mathbf{c} - \dfrac{\mathbf{b}}{2}$

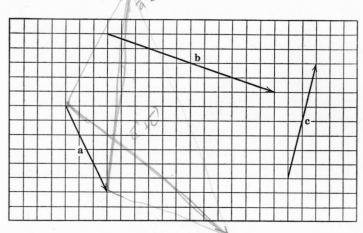

FIGURE #1-1

2. Copy the coplanar vectors given in Figure #1-2 and construct the vectors

   (a) $\mathbf{a} + 2\mathbf{b} - 3\mathbf{c}$ (b) $2\mathbf{a} - \dfrac{\mathbf{b} + 2\mathbf{c}}{3}$ (c) $m\mathbf{a}$ and $n\mathbf{b}$ such that $\mathbf{c} = m\mathbf{a} + n\mathbf{b}$

3. If $O(0, 0)$, $A(3, 5)$, $B(-1, 3)$, $C(1, -4)$ are points in a Cartesian coordinate plane and $\mathbf{a} = \overrightarrow{OA}$, $\mathbf{b} = \overrightarrow{OB}$, $\mathbf{c} = \overrightarrow{OC}$, construct the vectors

   (a) $\mathbf{a} + \mathbf{c}$ (b) $(\mathbf{a} + \mathbf{b}) + \mathbf{c}$ (c) $\mathbf{a} + 2\mathbf{c} - \mathbf{b}$

   (d) $(\mathbf{a} - \mathbf{b}) + (\mathbf{c} - \mathbf{b})$ (e) $\dfrac{\mathbf{b} + \mathbf{c}}{2} - \mathbf{a}$

4. If $A(2, 3)$, $B(5, 1)$, $C(0, -2)$, $D(-2, 2)$ are points in a Cartesian coordinate plane and $\mathbf{a} = \overrightarrow{AB}$, $\mathbf{b} = \overrightarrow{BC}$, $\mathbf{c} = \overrightarrow{CD}$, and $\mathbf{d} = \overrightarrow{DA}$, construct the vectors

   (a) $\mathbf{a} + \mathbf{c}$ (b) $\mathbf{a} - \mathbf{d}$ (c) $(\mathbf{a} + \mathbf{d}) - (\mathbf{c} + \mathbf{d})$
   (d) $(\mathbf{a} - \mathbf{c}) + (\mathbf{b} - \mathbf{d})$ (e) $\frac{1}{2}(\mathbf{c} - 3\mathbf{d})$

5. What are the magnitude and direction of the vector $\dfrac{\mathbf{n}}{|\mathbf{n}|}$?

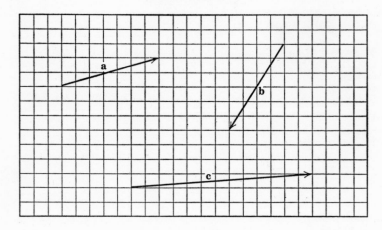

FIGURE #1-2

6. If $A(2, 3)$, $B(5, 7)$ are points in a Cartesian coordinate plane, what is the magnitude of $\overrightarrow{AB}$ and what is its direction with respect to the positive x-axis? What is the magnitude of $\dfrac{\overrightarrow{AB}}{|\overrightarrow{AB}|}$? $\left\|\dfrac{\overrightarrow{AB}}{|\overrightarrow{AB}|}\right\|.$

7. Let A, B, C, D be vertices of a tetrahedron. Let $\mathbf{b} = \overrightarrow{AB}$, $\mathbf{c} = \overrightarrow{AC}$, $\mathbf{d} = \overrightarrow{AD}$. Express the directed edges $\overrightarrow{BC}$, $\overrightarrow{BD}$, $\overrightarrow{CD}$ in terms of $\mathbf{b}$, $\mathbf{c}$, and $\mathbf{d}$.

8. Let A, B, C, D be vertices of a tetrahedron with O an arbitrary point not in the planes of the faces. Let $\mathbf{a} = \overrightarrow{OA}$, $\mathbf{b} = \overrightarrow{OB}$, $\mathbf{c} = \overrightarrow{OC}$, and $\mathbf{d} = \overrightarrow{OD}$. Express the directed edges $\overrightarrow{AB}$, $\overrightarrow{AC}$, and $\overrightarrow{AD}$ in terms of $\mathbf{a}$, $\mathbf{b}$, and $\mathbf{c}$.

9. Let A, B, C, and D be the vertices of a parallelogram $ABCD$. Let M be the point of intersection of the diagonals AC and BD. Let N be the midpoint of side AB. If $\mathbf{a} = \overrightarrow{AB}$ and $\mathbf{b} = \overrightarrow{BC}$, then find expressions for the vectors

   (a) $\overrightarrow{AC}$ (b) $\overrightarrow{AM}$ (c) $\overrightarrow{BD}$ (d) $\overrightarrow{BM}$

   (e) $\overrightarrow{AN}$ (f) $\overrightarrow{DN}$ (g) $\overrightarrow{MN}$

in terms of $\mathbf{a}$ and $\mathbf{b}$.

10. Given a cube with unit edges, find the magnitude of the sum of the three diagonal vectors drawn on the faces from one vertex of the cube.

11. Prove that for any two noncollinear vectors **a** and **b**

$$[x\mathbf{a} + y\mathbf{b} = z\mathbf{a} + w\mathbf{b}] \Leftrightarrow [x = z \text{ and } y = w]$$

12. Show by construction that
(a) $\mathbf{a} + \mathbf{b} = \mathbf{b} + \mathbf{a}$ (b) $m(\mathbf{a} + \mathbf{b}) = m\mathbf{a} + m\mathbf{b}$

13. Show graphically that
(a) $|\mathbf{a}| + |\mathbf{b}| \geq |\mathbf{a} + \mathbf{b}|$ (b) $|\mathbf{a}| - |\mathbf{b}| \leq |\mathbf{a} - \mathbf{b}|$

14. Prove Theorems 1.3.4 and 1.3.6.

B. Geometric Applications

(Use vector methods in the following problems unless otherwise stated.)

1. Do Example 1.2.2 with a different vector assignment than that used in the two solutions given in the text.

2. Let P be an arbitrary point and the triangle ABC be given. Prove that $|\overrightarrow{PA} + \overrightarrow{PB} + \overrightarrow{PC}| = |\overrightarrow{PM_1} + \overrightarrow{PM_2} + \overrightarrow{PM_3}|$, where M_1, M_2, and M_3 are the midpoints to the respective sides of the triangle.

3. Let a point P divide a segment AB so that $\dfrac{PB}{AP} = \dfrac{\alpha}{\beta}$. Show that for any point O, $\overrightarrow{OP} = \dfrac{\alpha}{\alpha + \beta}\mathbf{a} + \dfrac{\beta}{\alpha + \beta}\mathbf{b}$, where $\mathbf{a} = \overrightarrow{OA}$ and $\mathbf{b} = \overrightarrow{OB}$.

(*Note:* This is a generalization of Example 1.2.1.)

4. Prove that the line segment joining the midpoints of two sides of a triangle is parallel to and has a length one half that of the third side.

5. Prove that a line joining one vertex of a parallelogram to the midpoint of an opposite side trisects a diagonal of the parallelogram.

6. Prove that the medians of a triangle intersect at a point two thirds of the way from each vertex to its opposite side.

7. Let $ABCD$ be an arbitrary quadrilateral in 2-dimensional space. Let M_1, M_2, M_3, and M_4 be successive midpoints of its sides. Prove that M_1M_3 and M_2M_4 bisect each other. What kind of a figure is formed when M_1, M_2, M_3, and M_4 are connected by straight line segments?

8. Let $ABCD$ be an arbitrary quadrilateral in 3-dimensional space (not necessarily planar). Let M_1, M_2, M_3, and M_4 be successive midpoints of its sides. Prove that M_1M_3 and M_2M_4 meet and bisect each other. What kind of a figure is formed when M_1, M_2, M_3, and M_4 are connected by straight line segments?

Preface to problems 9 through 18

Vector equations may often be useful in describing loci. Recall that

DEFINITION *A locus is a collection of all points that satisfy one or more given conditions and which contains no point that does not satisfy these conditions.*

9. Let O be a fixed point and P be a variable point. What is the locus of all points P such that

(a) $|\overrightarrow{OP}| = 2$ (b) $|\overrightarrow{OP}| < 2$ (c) $|\overrightarrow{OP}| \leq 2$

(d) $|\overrightarrow{OP}| > 2$? (*Note:* Answer each part for 1-space, 2-space, and 3-space.)

10. Given point O and a positive scalar $k > 1$, find the locus of all points Q such that $|\overrightarrow{OP}| = k$ and $|\overrightarrow{PQ}| \leq 1$. Find the locus for 1-space, 2-space, and 3-space.

11. Let A and B be fixed points in a plane. Find the locus of all points P in the plane such that

(a) $|\overrightarrow{AP}| = |\overrightarrow{BP}|$ (b) $|\overrightarrow{AP}| + |\overrightarrow{BP}| = 2|\overrightarrow{AB}|$

(c) $|\overrightarrow{AP}| - |\overrightarrow{BP}| = 2|\overrightarrow{AB}|$

12. Given triangle ABC with P any point on BC. Find the locus of all points Q such that $\overrightarrow{PQ} = \overrightarrow{AP} + \overrightarrow{PB} + \overrightarrow{PC}$.

13. If O, A, and B are three fixed points not all on one line, then find a vector equation for the line determined by A and B in terms of the vectors $\overrightarrow{OA}$ and $\overrightarrow{OB}$.

14. Find a vector equation for the bisector of the angle between two intersecting lines. (*Hint:* Let **a** and **b** be vectors on the lines. Note that there are two bisectors.)

15. Given O, A, and B not all on one line, find a vector equation defining the locus of all points P on the plane determined by O, A, and B.

16. Let A and B be given points. Find a vector equation of the locus of all points P such that

(a) P is on the line segment defined by AB.

(b) P is on the plane perpendicular to AB through A.

(c) P is on the lateral surface of a right circular cylinder with axis AB and radius 1 unit.

17. Given points A and B, find the equation of the locus of all points P lying on a circle with diameter AB. (*Hint:* Recall that a triangle inscribed in a semicircle is a right triangle.)

18. Given points A and B, find the equation of the locus of all points P lying on a sphere with diameter AB.

C. Physics Applications

1. Classify the following into vector or scalar quantities:
   (a) velocity   (b) density   (c) temperature
   (d) directed line segment   (e) mass   (f) force

2. Classify the following into vector or scalar quantities:
   (a) speed   (b) acceleration   (c) length
   (d) displacement   (e) pressure   (f) wind velocity

3. An airplane flies East for a distance of 10 miles, then turns and flies northeast for a distance of 5 miles. Construct the resultant displacement vector and state its direction and magnitude.

4. A ship travels successively 5 miles southeast, 3 miles northeast, and 4 miles North. Construct the resultant displacement vector and state its direction and magnitude.

5. A plane, after flying for $2\frac{1}{2}$ hours at 800 mph in a direction which is 30° east of North, is forced down. Determine graphically and also by calculation how far North and how far East the plane is from its initial starting point.

6. A plane, after flying for $3\frac{1}{4}$ hours at 1200 mph in a direction which is 60° south of West, is forced down. Find the coordinates of its position if the positive y-axis is in the direction of North and the positive x-axis is pointed to the East, with the origin of the coordinate system at the starting point of the plane.

Preface to problems 7 through 10

Study the solution of the following illustrative example before working problems 7 through 10.

A boat capable of traveling 12 miles per hour on still water maintains a westward compass reading in crossing a river. If the river is flowing southward at 4 miles per hour, what is the velocity of the boat with respect to the land? (Figure #1-3.)

Solution

(i) Construct separate vectors representing the boat's motion and the river's flow, respectively. Label the terminal ends as shown and the initial ends with reference to the fixed element.

(ii) Form the vector resultant by placing the "like" labels together and drawing the vector represented by the segment connecting the "unlike" labels. The vector $\overrightarrow{LB}$ represents the velocity of the boat with respect to the land. Its magnitude is 12.65 mph and its direction 18°26′ south of West. (Figure #1-4.)

7. A plane, after flying for 3 hours at 400 mph in a compass direction of 30° north of West, is forced down. If there has been a steady cross wind of 60 mph in a northward direction, find the location of the plane with respect to its starting point. First determine the answer graphically using protractor and ruler, then calculate the location algebraicly, using the necessary trigonometric functions.

8. A plane with a cruising speed of 300 mph maintains a compass heading of 30° west of North. A tailwind of 40 mph is coming from the Southeast. Determine graphically and algebraicly the velocity of the airplane with respect to the ground. (Remember that velocity is a vector quantity.)

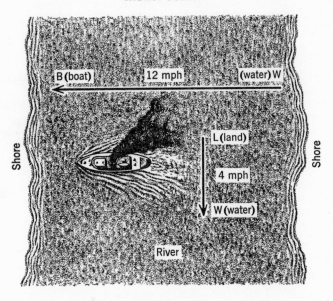

FIGURE #1-3

9. Two piers are directly opposite each other on a river that flows South. A man wishes to go (in a straight line) from one pier to the other in a boat with a cruising speed of 8 knots. If the river's current is 2.5 knots, what compass heading should be maintained while crossing the river? What is the actual velocity of the boat with respect to the land?

10. A ship is sailing due South at a speed of 20 knots with respect to the land. An 8-knot wind blows from the West. What angle will a smoke screen sent out by the ship make with the ship's course? How long will the smoke screen be after 20 minutes of generation?

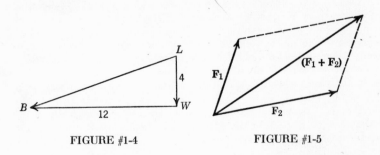

FIGURE #1-4   FIGURE #1-5

Preface to Problems 11 through 16

Forces are vector quantities that are subject to the rules of vector addition; that is, two forces F_1 and F_2 acting on a body are equivalent to the single force $(F_1 + F_2)$ acting on that body. (Figure #1-5.)

11. Construct the following vectors graphically:
(*Note:* Indicate the scale of measurement used.)

    (a) A force of 12 lb in a direction 30° south of East.

    (b) A force of 15 lb in a direction 45° west of North.

    (c) The sum of the forces given in (a) and (b).

    (d) The difference of the forces given in (a) and (b).

12. Construct the following vectors graphically:
(*Note:* All angles are measured in a counterclockwise direction with respect to the positive x-axis.)

    (a) A force F_1 of 350 lb at 60°.

    (b) A force F_2 of 200 lb at 120°.

    (c) A force F_3 of 500 lb at 210°.

    (d) $F_4 = F_1 + F_2 + F_3$.

    (e) $F_5 = (F_1 + F_2) - F_3$.

13. Three forces act simultaneously on the same object; 10 gm at 30°, 20 gm at 330°, and 15 $\sqrt{3}$ gm at 180° where the angles are measured in a counterclockwise direction with respect to the positive x-axis. Find their resultant graphically and also by calculation.

14. Two forces act on the same object: 25 dynes at 40° and 43 dynes at 100°. Find the resultant of the forces both graphically and by calculation.

15. A force of 76 lb acts at an angle of 60° on an object located at the origin. What two forces in the x- and y-axis directions, respectively, will have an equivalent effect on the object?

16. A force of 200 lb acts at an angle of 150° on an object located at the origin. What two forces in the x- and y-axis directions, respectively, will have an equivalent effect on the object?

2

the scalar product
base vectors

2.1 THE SCALAR (DOT) PRODUCT

In Section 1.2 the multiplication of a vector by a scalar was defined (Definition 1.2.3). There are two further common types of products involving vectors: "scalar products" and "vector products." In this section we will define and discuss the notion of a scalar product. The notion of a vector product will be developed in Chapter 3. The reasons for having two distinct notions for the product of vectors will become apparent as new concepts are developed and new applications considered.

DEFINITION 2.1.1 *The* **angle between two** *arbitrary* **vectors a** *and* **b**, *denoted by* θ *or* $\measuredangle(\mathbf{a}, \mathbf{b})$, *is the angle between the two vectors when taken from a common initial point. The angle θ is restricted to the interval $0 \leq \theta \leq \pi$.*

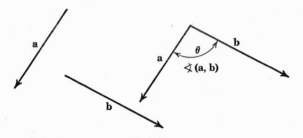

FIGURE 2.1.1   The angle between two vectors.

The quantities $|a| \cos \theta$, $|b| \cos \theta$, and $|a| |b| \cos \theta$ arise frequently in applications.   Thus it is useful to give these quantities names and to observe a few of their special properties.

DEFINITION 2.1.2 *Let vectors* **a** *and* **b** *be drawn from a common initial point with* $\theta = \angle (a, b)$. *The* **component of a in the direction of b,** *denoted by* $\text{comp}_b \, a$, *and the component of* **b** *in the direction of* **a,** *denoted by* $\text{comp}_a \, b$, *are defined by the following:*

$$\text{(i)} \quad \text{comp}_b \, a = |a| \cos \theta$$
$$\text{(ii)} \quad \text{comp}_a \, b = |b| \cos \theta *$$

We observe the following:
If $0 \leq \theta < \pi/2$, then $\text{comp}_b \, a$ and $\text{comp}_a \, b$ are positive.
If $\theta = \pi/2$, then $\text{comp}_b \, a = \text{comp}_a \, b = 0$.
If $\pi/2 < \theta \leq \pi$, then $\text{comp}_b \, a$ and $\text{comp}_a \, b$ are negative.

THEOREM 2.1.1 Given any three vectors **a**, **b**, and **c**, then $\text{comp}_a \, (b + c) = \text{comp}_a \, b + \text{comp}_a \, c$.

A method for the proof of this theorem is suggested by Figure 2.1.3.

DEFINITION 2.1.3 **The scalar product** (*also called the "dot" or "inner" product) of two vectors, denoted by* **a** · **b,** *is the scalar quantity given by:* $a \cdot b = |a| |b| \cos \theta$ *where* $\theta = \angle (a, b).$†

* Note that $\text{comp}_b \, a$ and $\text{comp}_a \, b$ are both scalar quantities.
† Note that it does not matter whether θ is taken positive or negative, for $\cos (-\theta) = \cos \theta$.

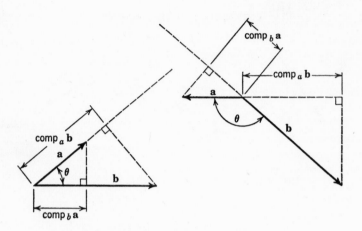

FIGURE 2.1.2 comp_b **a** and comp_a **b**.

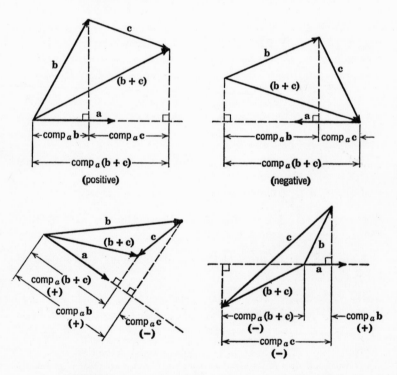

FIGURE 2.1.3 comp_a (**b** + **c**) = comp_a **b** + comp_a **c**.

24

The notion of the scalar product has its historical roots in physics, where the concept arises rather naturally. It has also been found to be of considerable use in many other fields. It will play an important role in much of the discussion and development that follow.

Observe that the definition of the scalar product is independent of any coordinate system. The basis for the use of the scalar product in many applications is given in the following two theorems.

THEOREM 2.1.2 $\mathbf{a} \cdot \mathbf{b} = a(\text{comp}_a \mathbf{b}) = b(\text{comp}_b \mathbf{a})$

THEOREM 2.1.3   If $\mathbf{a}$ and $\mathbf{b}$ are nonzero vectors, then

(i) $[\mathbf{a} \cdot \mathbf{b} > 0] \Leftrightarrow \left[0 \leq \measuredangle (\mathbf{a}, \mathbf{b}) < \dfrac{\pi}{2} \right]$

(ii) $[\mathbf{a} \cdot \mathbf{b} = 0] \Leftrightarrow [\mathbf{a} \text{ perpendicular to } \mathbf{b}]$

(iii) $[\mathbf{a} \cdot \mathbf{b} < 0] \Leftrightarrow \left[\dfrac{\pi}{2} < \measuredangle (\mathbf{a}, \mathbf{b}) \leq \pi \right]$

Theorem 2.1.2 follows immediately from Definition 2.1.2 and 2.1.3; the proof of Theorem 2.1.3 will be left as an exercise for the reader. Part (ii) of Theorem 2.1.3 will be found to be a particularly useful property of the scalar product.

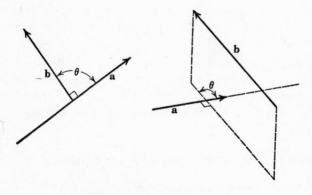

FIGURE 2.1.4 $\mathbf{a} \cdot \mathbf{b} = 0$.

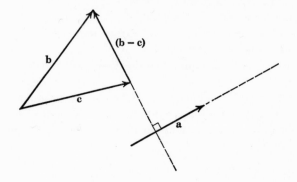

FIGURE 2.1.5 $\mathbf{a} \cdot \mathbf{b} = \mathbf{a} \cdot \mathbf{c}, \mathbf{b} \neq \mathbf{c}.$

Algebraic operations involving the scalar product are based on the following laws.

ALGEBRAIC LAWS ($\mathbf{a}$, $\mathbf{b}$, and $\mathbf{c}$ are vectors; h is a scalar)

2.1.1	$\mathbf{a} \cdot \mathbf{b} = \mathbf{b} \cdot \mathbf{a}$	commutative law
2.1.2	$\mathbf{a} \cdot (\mathbf{b} + \mathbf{c}) = \mathbf{a} \cdot \mathbf{b} + \mathbf{a} \cdot \mathbf{c}$	distributive law
2.1.3	$\mathbf{a} \cdot (h\mathbf{b}) = (h\mathbf{a}) \cdot \mathbf{b} = h(\mathbf{a} \cdot \mathbf{b})$	associative law
2.1.4	$\mathbf{a} \cdot \mathbf{a} = |\mathbf{a}|^2 = a^2$	

Laws 2.1.1, 2.1.3, and 2.1.4 are direct consequences of the definition of $\mathbf{a} \cdot \mathbf{b}$.   The distributive law is proved using Theorems 2.1.2 and 2.1.1.

If $\mathbf{a} \cdot \mathbf{b} = \mathbf{a} \cdot \mathbf{c}$, we might ask, "does $\mathbf{b}$ necessarily equal $\mathbf{c}$?" That is, does the "cancellation law" hold for vector quantities? The "cancellation law" does not hold for vector quantities. Suppose $\mathbf{a} \cdot \mathbf{b} = \mathbf{a} \cdot \mathbf{c}$; then

(i) $\mathbf{a} \cdot \mathbf{b} - \mathbf{a} \cdot \mathbf{c} = 0$ (Definition 1.2.2)

(ii) $\mathbf{a} \cdot (\mathbf{b} - \mathbf{c}) = 0$ (Algebraic Law 2.1.2)

(iii) $\mathbf{a}$ is perpendicular to $(\mathbf{b} - \mathbf{c})$ (Theorem 2.1.3)

(iv) Hence we see that $\mathbf{a} \cdot \mathbf{b}$ can equal $\mathbf{a} \cdot \mathbf{c}$ without $\mathbf{b}$ being equal to $\mathbf{c}$.   (Figure 2.1.5.)

2.2   APPLICATIONS OF THE SCALAR PRODUCT

The notion of projections or components is useful in many applications.   For example, the work done by a constant force on a moving object is defined in physics to be the product of the component of the force in the direction of motion of the object and the distance the object is displaced, that is,

Work done = (component of force in direction of motion)

$\times$ (displacement of object)

$$= (\text{comp}_{AB} \, \mathbf{F})(|\overrightarrow{AB}|)$$
$$= |\mathbf{F}||\overrightarrow{AB}|[\cos \measuredangle(\mathbf{F}, \overrightarrow{AB})]$$
$$= \mathbf{F} \cdot \overrightarrow{AB}$$

EXAMPLE 2.2.1 If a force of 10 lb is applied at 30° to the direction of motion of an object and the object is moved 12 ft, find the work done by the force.

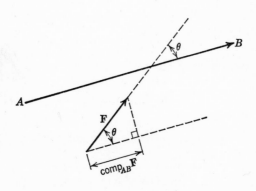

FIGURE 2.2.1   Work done $= \mathbf{F} \cdot \overrightarrow{AB}$.

Solution

$$\text{Work done} = \mathbf{F} \cdot \overrightarrow{AB}$$
$$= (10)(12) \cos 30°$$
$$= 60 \sqrt{3}$$
$$= 103.9 \text{ ft lb (approx.)}$$

One of the important applications of scalar products, as mentioned earlier, is their use to express perpendicularity between two vectors.

EXAMPLE 2.2.2   Show that the angle inscribed in a semicircle is a right angle.

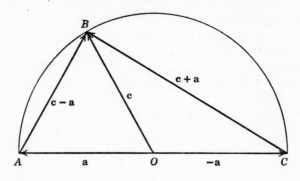

FIGURE 2.2.2

Solution

Let vectors be assigned as indicated in Figure 2.2.2.   We then have
(i) $|\mathbf{a}| = |\mathbf{c}| = |-\mathbf{a}| =$ radius of the circle (r).

(ii)
$$\begin{aligned}
\overrightarrow{AB} \cdot \overrightarrow{CB} &= (\mathbf{c} - \mathbf{a}) \cdot (\mathbf{c} + \mathbf{a}) \\
&= \mathbf{c} \cdot \mathbf{c} - \mathbf{a} \cdot \mathbf{a} \\
&= c^2 - a^2 \\
&= r^2 - r^2 \\
&= 0
\end{aligned}$$

(iii) Therefore $\overrightarrow{AB}$ is perpendicular to $\overrightarrow{CB}$ and $\measuredangle ABC$ is a right angle.

Another application of the scalar product is illustrated in the following derivation of the law of cosines for triangles.

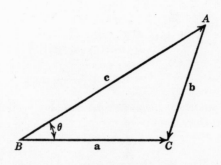

FIGURE 2.2.3

EXAMPLE 2.2.3   Prove the law of cosines for plane triangles.

Solution

Let ABC be an arbitrary triangle with vectors assigned as shown in Figure 2.2.3. We then have

(i) $\mathbf{b} = \mathbf{a} - \mathbf{c}$
(ii) $\mathbf{b} \cdot \mathbf{b} = (\mathbf{a} - \mathbf{c}) \cdot (\mathbf{a} - \mathbf{c})$
(iii) $\mathbf{b} \cdot \mathbf{b} = \mathbf{a} \cdot \mathbf{a} + \mathbf{c} \cdot \mathbf{c} - 2(\mathbf{a} \cdot \mathbf{c})$
(iv) Therefore $b^2 = a^2 + c^2 - 2ac \cos \theta$

SUMMARY

2.1 The Scalar Product

DEFINITION 2.1.1   *The* **angle between two arbitrary vectors a** *and* **b** *is the angle between the two vectors when taken from a common initial point.   The angle θ is restricted to the interval $0 \leq \theta \leq \pi$.*

DEFINITION 2.1.2   *The* **component of a in the direction of b** *is defined by the following:*

$$\text{comp}_b \, \mathbf{a} = |\mathbf{a}| \cos \theta$$

THEOREM 2.1.1   Given any three vectors **a**, **b**, and **c**, then

$$\text{comp}_a \, (\mathbf{b} + \mathbf{c}) = \text{comp}_a \, \mathbf{b} + \text{comp}_a \, \mathbf{c}$$

DEFINITION 2.1.3   *The* **scalar product** *of two vectors* **a** *and* **b** *is the scalar quantity given by*

$$\mathbf{a} \cdot \mathbf{b} = |\mathbf{a}| \, |\mathbf{b}| \cos \theta \text{ where } \theta = \sphericalangle (\mathbf{a}, \mathbf{b})$$

THEOREM 2.1.2 $\mathbf{a} \cdot \mathbf{b} = a(\text{comp}_a \, \mathbf{b}) = b(\text{comp}_b \, \mathbf{a})$

THEOREM 2.1.3   If **a** and **b** are nonzero vectors, then

(i) $[\mathbf{a} \cdot \mathbf{b} > 0] \Leftrightarrow \left[0 \leq \sphericalangle (\mathbf{a}, \mathbf{b}) < \dfrac{\pi}{2} \right]$

(ii) $[\mathbf{a} \cdot \mathbf{b} = 0] \Leftrightarrow [\mathbf{a} \text{ perpendicular to } \mathbf{b}]$

(iii) $[\mathbf{a} \cdot \mathbf{b} < 0] \Leftrightarrow \left[\dfrac{\pi}{2} < \sphericalangle (\mathbf{a}, \mathbf{b}) \leq \pi \right]$

ALGEBRAIC LAWS (**a**, **b**, and **c** are vectors; h is a scalar)

2.1.1	$\mathbf{a} \cdot \mathbf{b} = \mathbf{b} \cdot \mathbf{a}$	commutative law
2.1.2	$\mathbf{a} \cdot (\mathbf{b} + \mathbf{c}) = \mathbf{a} \cdot \mathbf{b} + \mathbf{a} \cdot \mathbf{c}$	distributive law
2.1.3	$\mathbf{a} \cdot (h\mathbf{b}) = (h\mathbf{a}) \cdot \mathbf{b} = h(\mathbf{a} \cdot \mathbf{b})$	associative law
2.1.4	$\mathbf{a} \cdot \mathbf{a} = |\mathbf{a}|^2 = a^2$	

2.2 Applications of the Scalar Product

EXAMPLE 2.2.1 Work = $\mathbf{F} \cdot \overrightarrow{AB}$.
EXAMPLE 2.2.2 Geometric: angle inscribed in a semicircle.
EXAMPLE 2.2.3 Trigonometric: law of cosines.

PROBLEM SET #2.1

A. General

1. Prove Theorem 2.1.2.

2. Prove Theorem 2.1.3.

3. Prove Algebraic Law 2.1.1.

4. Prove Algebraic Law 2.1.2.

5. Prove Algebraic Law 2.1.3.

6. Prove Algebraic Law 2.1.4.

7. Show that $(\mathbf{a} + \mathbf{b}) \cdot (\mathbf{c} + \mathbf{d}) = \mathbf{a} \cdot \mathbf{c} + \mathbf{a} \cdot \mathbf{d} + \mathbf{b} \cdot \mathbf{c} + \mathbf{b} \cdot \mathbf{d}$. Justify each step with a definition, law, or theorem.

8. Write a formula for the magnitude of a vector $\mathbf{a}$ in terms of the scalar product.

9. If $\mathbf{i}$, $\mathbf{j}$, and $\mathbf{k}$ are three vectors such that $|\mathbf{i}| = |\mathbf{j}| = |\mathbf{k}| = 1$ and $\mathbf{i} \cdot \mathbf{j} = \mathbf{j} \cdot \mathbf{k} = \mathbf{k} \cdot \mathbf{i} = 0$, then show that for any vector $\mathbf{v}$ in 3-space $\mathbf{v} = (\mathbf{v} \cdot \mathbf{i})\mathbf{i} + (\mathbf{v} \cdot \mathbf{j})\mathbf{j} + (\mathbf{v} \cdot \mathbf{k})\mathbf{k}$. (*Hint:* See Figure 1.3.2.)

10. Prove that $(\mathbf{a} \cdot \mathbf{b})(\mathbf{a} \cdot \mathbf{b}) \leq a^2 b^2$. Under what conditions will the equality hold?

B. Geometry

1. Prove that the median to the base of an isosceles triangle is perpendicular to the base.

2. Prove that the diagonals of a rhombus are perpendicular to each other.

3. Prove that the sum of the squares of the diagonals of a parallelogram is equal to the sum of the squares of its sides. (*Hint:* $a^2 = \mathbf{a} \cdot \mathbf{a}$.)

4. Prove that the altitudes of a triangle are concurrent.

5. Prove that the perpendicular bisectors of a triangle are concurrent.

6. Let A, B, and C be three noncollinear points. Write a vector formula for the distance between the point C and the line connecting A and B.

7. Given the fixed points O and C, find the locus of all points P such that $(\overrightarrow{OC} - \overrightarrow{OP}) \cdot \overrightarrow{OC} = 0$.

8. Given points O, C, and a scalar $r > 0$, find the locus of all points P such that $(\overrightarrow{OP} - \overrightarrow{OC})^2 = r^2$.

C. Physics

1. An object is displaced along a straight line from $A(1, 0)$ to $B(3, 2)$ while being acted upon by a constant force of 10 lb in the positive x-axis direction. Assuming the units are in feet, find the work done by this force in displacing the object from A to B.

2. Find the total work done by the force in Problem 1 if the object is moved along the broken line path from $A(1, 0)$ to $B(3, 2)$ then to $C(3, 2\sqrt{3})$.

3. Find the total work done by the force in Problem 1 if the object is moved completely around the triangle (in either direction) with vertices $A(1, 0)$, $B(3, 2)$, $C(3, 2\sqrt{3})$.

4. An object is pushed 6 ft along a table by a 20-lb force parallel to the table. Friction creates an opposing force of 3 lb.
   (a) How much work is done by the 20-lb force?
   (b) How much work is done by the friction force?
   (c) What is the total work of the forces combined?
   (d) If the 20-lb force acts downward at an angle of 30° to the table, then answer parts (a), (b), and (c).

5. A downward force of 100 lb making an angle of 60° with a table, pushes a 50 lb object 5 ft. The coefficient of friction between the object and the table is $\alpha = 0.1$. What is the total work done by the combined forces?

6. An upward force of 50 lb, making an angle of 30° with the top of a table, acts on a 100-lb object so as to move it 10 ft. The coefficient of

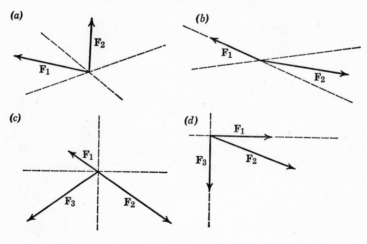

FIGURE #2.1-1

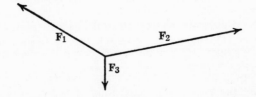

FIGURE #2.1-2

friction between the object and the table is $\alpha = 0.2$.   What is the total work done by the combined forces?

7. Three forces $\mathbf{F}_1$, $\mathbf{F}_2$, $\mathbf{F}_3$ are applied at a point P which is conveniently selected as the origin of a coordinate system.   The vectors then have their initial ends at $(0, 0)$ and their terminal ends at $(2, -3)$, $(5, 7)$, and $(-3, -2)$, respectively.   Find both graphically and by calculation two forces (vectors) $\overrightarrow{OX}$ and $\overrightarrow{OY}$ parallel to the x- and y-axes which would have an equivalent effect as the total of $\mathbf{F}_1$, $\mathbf{F}_2$, $\mathbf{F}_3$.   (*Hint:* Use the components of $\mathbf{F}_1$, $\mathbf{F}_2$, and $\mathbf{F}_3$ in the positive x and y directions.)

8. Two forces $\mathbf{F}_1$ and $\mathbf{F}_2$ act along nonintersecting lines.   Show that their resultant is perpendicular to the common perpendicular to the two lines.

9. One of the (two) requirements for static equilibrium of a body is to have the sum of all the forces acting on the body equal to zero.   Copy Figure 2.1-1 and, using component notions, graphically find the forces required along the given lines to obtain this condition of equilibrium.

10. Find a single force $\mathbf{u}$ such that $\mathbf{F}_1 + \mathbf{F}_2 + \mathbf{F}_3 + \mathbf{u} = 0$ in Figure 2.1-2.

2.3   BASE VECTORS AND CARTESIAN COORDINATE SYSTEMS

It was noted earlier that many problems can be stated and solved using vector quantities free of any coordinate system. There are times, however, when a coordinate system is useful. In the development that follows we will be dealing with ordinary Euclidean space of two and three dimensions.

For 3-dimensional space we will choose a "right-handed" coordinate system.   A right-handed system is obtained if, when we rotate the positive x-axis through 90° in a counter-clockwise manner, the forward motion of a right-handed screw will be in the positive z-axis direction. Another method for describing a right-handed system is to point the thumb of the right hand in the direction of the positive z-axis and to note whether the fingers, when relaxed, point in the direction of rotation which carries the positive x-axis into the positive y-axis through an angle of less than 180°.

For a left-handed system the forward motion of a left-hand screw would be in the positive z-axis direction. Alternatively, the left hand would be used to orient the axes.

In general, it is desirable to use the same orientation of a coordinate system consistently, for certain formulas are changed with a change in the orientation.

The reader should review Section 1.3 at this time before proceeding further. Theorems 1.3.1 and 1.3.2 are of particular importance.

DEFINITION 2.3.1   [**a** *is a unit vector*] $\Leftrightarrow$ [$|\mathbf{a}| = 1$]

DEFINITION 2.3.2   *If every vector in a given space can be represented uniquely as a linear combination of a given set S of vectors in that space, then S is called a* **basis** *of the set of all vectors in that space and the elements of S are called* **base vectors.**

An immediate consequence of Definition 2.3.2 and Theorems 1.3.1 and 1.3.2 is that any three (nonzero) noncoplanar vectors in 3-space form a basis for all vectors in the 3-space.   Also, any

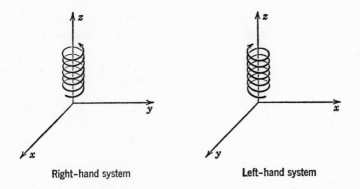

Right-hand system   Left-hand system

FIGURE 2.3.1   Orientation of coordinates.

two (nonzero) noncollinear vectors in 2-space form a basis for all the vectors in the 2-space.*

A very simple and elegant basis for the vectors in 3-space are the three unit vectors defined as follows:

DEFINITION 2.3.3   *Given the four points $O(0, 0, 0)$, $P_1(1, 0, 0)$, $P_2(0, 1, 0)$, $P_3(0, 0, 1)$ in a Cartesian coordinate space, we define*

$$\mathbf{i} = \overrightarrow{OP}_1, \quad \mathbf{j} = \overrightarrow{OP}_2, \quad \mathbf{k} = \overrightarrow{OP}_3$$

The reader should have little difficulty in establishing the following theorem and its corollary.

THEOREM 2.3.1   (i) $\mathbf{i} \cdot \mathbf{i} = \mathbf{j} \cdot \mathbf{j} = \mathbf{k} \cdot \mathbf{k} = 1$
(ii) $\mathbf{i} \cdot \mathbf{j} = \mathbf{j} \cdot \mathbf{k} = \mathbf{k} \cdot \mathbf{i} = 0$

COROLLARY 2.3.1   The vectors $\mathbf{i}$, $\mathbf{j}$, and $\mathbf{k}$ form a basis for all vectors in 3-space.

Since $\mathbf{i}$, $\mathbf{j}$, and $\mathbf{k}$ form a basis for the vectors in 3-space, every vector $\mathbf{a}$ in the 3-space can be uniquely represented as a linear combination of these base vectors, that is, for each vector $\mathbf{a}$ there is a unique set of scalars a_x, a_y, a_z such that $\mathbf{a} = a_x\mathbf{i} + a_y\mathbf{j} + a_z\mathbf{k}$.

THEOREM 2.3.2 $a_x = \text{comp}_i\, \mathbf{a} = \mathbf{a} \cdot \mathbf{i}$
$a_y = \text{comp}_j\, \mathbf{a} = \mathbf{a} \cdot \mathbf{j}$
$a_z = \text{comp}_k\, \mathbf{a} = \mathbf{a} \cdot \mathbf{k}$

Noting that $\mathbf{i}$, $\mathbf{j}$, and $\mathbf{k}$ are unit vectors, the reader should readily be able to establish the above theorem as a consequence of Theorem 2.1.2. We call the scalars a_x, a_y, and a_z the **components** of $\mathbf{a}$ in the x-direction, the y-direction, and the z-direction, respectively. These scalars are also often called the "$\mathbf{i}$," "$\mathbf{j}$," and "$\mathbf{k}$" components of $\mathbf{a}$.

The components a_x, a_y, and a_z have the following properties:

THEOREM 2.3.3 $\mathbf{a} + \mathbf{b} = (a_x + b_x)\mathbf{i} + (a_y + b_y)\mathbf{j} + (a_z + b_z)\mathbf{k}$

THEOREM 2.3.4 $\mathbf{a} - \mathbf{b} = (a_x - b_x)\mathbf{i} + (a_y - b_y)\mathbf{j} + (a_z - b_z)\mathbf{k}$

---

* In general, any set of n nonzero linearly independent vectors in an n-dimensional vector space forms a basis for that space.

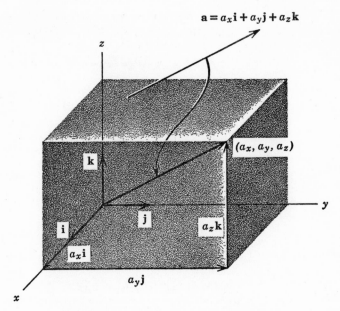

FIGURE 2.3.2   Base vectors and components.

Proof $\mathbf{a} - \mathbf{b} = (a_x\mathbf{i} + a_y\mathbf{j} + a_z\mathbf{k}) - (b_x\mathbf{i} + b_y\mathbf{j} + b_z\mathbf{k})$
$= a_x\mathbf{i} + a_y\mathbf{j} + a_z\mathbf{k} - b_x\mathbf{i} - b_y\mathbf{j} - b_z\mathbf{k}$
$= (a_x\mathbf{i} - b_x\mathbf{i}) + (a_y\mathbf{j} - b_y\mathbf{j}) + (a_z\mathbf{k} - b_z\mathbf{k})$
$= (a_x - b_x)\mathbf{i} + (a_y - b_y)\mathbf{j} + (a_z - b_z)\mathbf{k}$

Note the use of Algebraic Laws 1.2.1–1.2.9.

THEOREM 2.3.5 $h\mathbf{a} = (ha_x)\mathbf{i} + (ha_y)\mathbf{j} + (ha_z)\mathbf{k}$

THEOREM 2.3.6 $[\mathbf{a} = \mathbf{0}] \Leftrightarrow [a_x = 0, a_y = 0, a_z = 0]$

Proof   1. $[\mathbf{a} = \mathbf{0}] \Rightarrow [a_x = 0, a_y = 0, a_z = 0]$
$a_x = \mathbf{a} \cdot \mathbf{i} = \mathbf{0} \cdot \mathbf{i} = 0(1) \cos \alpha = 0$
$a_y = \mathbf{a} \cdot \mathbf{j} = \mathbf{0} \cdot \mathbf{j} = 0(1) \cos \beta = 0$
$a_z = \mathbf{a} \cdot \mathbf{k} = \mathbf{0} \cdot \mathbf{k} = 0(1) \cos \gamma = 0$
2. $[a_x = 0, a_y = 0, a_z = 0] \Rightarrow [\mathbf{a} = \mathbf{0}]$
$\mathbf{a} = a_x\mathbf{i} + a_y\mathbf{j} + a_z\mathbf{k} = 0\mathbf{i} + 0\mathbf{j} + 0\mathbf{k} = \mathbf{0} + \mathbf{0} + \mathbf{0} = \mathbf{0}$

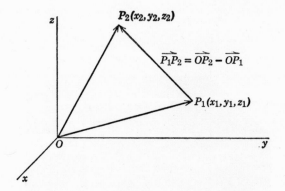

FIGURE 2.3.3

THEOREM 2.3.7 $[\mathbf{a} = \mathbf{b}] \Leftrightarrow [a_x = b_x, a_y = b_y, a_z = b_z]$

Proof 1. $[\mathbf{a} = \mathbf{b}] \Rightarrow [a_x = b_x, a_y = b_y, a_z = b_z]$
$\mathbf{a} - \mathbf{b} = \mathbf{0}$
$(a_x - b_x)\mathbf{i} + (a_y - b_y)\mathbf{j} + (a_z - b_z)\mathbf{k} = \mathbf{0}$
$a_x - b_x = 0, a_y - b_y = 0, a_z - b_z = 0$
$a_x = b_x, a_y = b_y, a_z = b_z$
    2. $[a_x = b_x, a_y = b_y, a_z = b_z] \Rightarrow [\mathbf{a} = \mathbf{b}]$

The steps of 1 are reversible.

THEOREM 2.3.8 $\mathbf{a} \cdot \mathbf{b} = a_x b_x + a_y b_y + a_z b_z$

Theorem 2.3.8, in conjunction with Definition 2.1.3 for the scalar product, is of particular importance in later applications. Note that the vector operation is evaluated in terms of the scalar components of the given vectors.

THEOREM 2.3.9 $\mathbf{a} \cdot \mathbf{a} = a_x^2 + a_y^2 + a_z^2 = a^2 = |\mathbf{a}|^2$ so that

$$|\mathbf{a}| = \sqrt{a_x^2 + a_y^2 + a_z^2}$$

THEOREM 2.3.10   Given the points $P_1(x_1, y_1, z_1)$ and $P_2(x_2, y_2, z_2)$, we have

$$\overrightarrow{P_1 P_2} = (x_2 - x_1)\mathbf{i} + (y_2 - y_1)\mathbf{j} + (z_2 - z_1)\mathbf{k}$$

Expressing the directed line segment from P_1 to P_2 as a vector destroys the geometric location of P_1P_2 (i.e., $\overrightarrow{P_1P_2}$ is in a free vector form).   Note, however, that

$$|\overrightarrow{P_1P_2}| = \sqrt{(x_2 - x_1)^2 + (y_2 - y_1)^2 + (z_2 - z_1)^2}$$

by Theorem 2.3.9.

A proof of Theorem 2.3.10 is suggested in Figure 2.3.3.

THEOREM 2.3.11   If $\theta = \measuredangle(\mathbf{a}, \mathbf{b})$, then

$$\cos\theta = \frac{\mathbf{a} \cdot \mathbf{b}}{|\mathbf{a}|\,|\mathbf{b}|} = \frac{a_xb_x + a_yb_y + a_zb_z}{\sqrt{a_x{}^2 + a_y{}^2 + a_z{}^2}\,\sqrt{b_x{}^2 + b_y{}^2 + b_z{}^2}}$$

THEOREM 2.3.12   If $\mathbf{a}$ is a nonzero vector, then

$$\frac{\mathbf{a}}{|\mathbf{a}|} = \frac{a_x}{|\mathbf{a}|}\,\mathbf{i} + \frac{a_y}{|\mathbf{a}|}\,\mathbf{j} + \frac{a_z}{|\mathbf{a}|}\,\mathbf{k}$$

is a unit vector pointing in the direction of $\mathbf{a}$.

2.4   DIRECTION COSINES AND DIRECTION NUMBERS

In working with vectors we note that the components of any nonzero vector are directly proportional to the cosines of the angles that the vector makes with the positive coordinate axes. This observation is formalized to provide a useful tool in solving problems involving lines, planes, and/or surfaces.

DEFINITION 2.4.1   *The* **direction angles** *of a vector* **a** *are the angles* $\alpha = \measuredangle(\mathbf{a}, \mathbf{i})$, $\beta = \measuredangle(\mathbf{a}, \mathbf{j})$, $\gamma = \measuredangle(\mathbf{a}, \mathbf{k})$. *The cosines of these angles,* $\cos\alpha$, $\cos\beta$, $\cos\gamma$, *are called the* **direction cosines** *of the vector* **a**.

If we note that $\cos\alpha = \dfrac{\mathbf{a} \cdot \mathbf{i}}{|\mathbf{a}|\,|\mathbf{i}|}$ (Theorem 2.3.11) and expand this expression, using Theorem 2.3.12, we obtain $\cos\alpha = \dfrac{a_x}{|\mathbf{a}|}$; similarly, $\cos\beta = \dfrac{a_y}{|\mathbf{a}|}$ and $\cos\gamma = \dfrac{a_z}{|\mathbf{a}|}$.   Hence we have established Theorem 2.4.1.

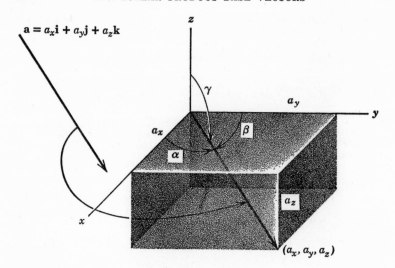

$$\mathbf{a} = a_x\mathbf{i} + a_y\mathbf{j} + a_z\mathbf{k}$$

FIGURE 2.4.1   Direction angles and components.

THEOREM 2.4.1   The direction cosines of a vector $\mathbf{a}$ are

$$\cos \alpha = \frac{a_x}{|\mathbf{a}|}, \qquad \cos \beta = \frac{a_y}{|\mathbf{a}|}, \qquad \cos \gamma = \frac{a_z}{|\mathbf{a}|}$$

COROLLARY 2.4.1   The components of a unit vector having the same direction as $\mathbf{a}$ are the direction cosines of $\mathbf{a}$.

COROLLARY 2.4.2   If $\cos \alpha$, $\cos \beta$, $\cos \gamma$ are the direction cosines of a vector $\mathbf{a}$, then

$$(\cos \alpha)^2 + (\cos \beta)^2 + (\cos \gamma)^2 = 1.$$

DEFINITION 2.4.2   *A set (triple) of numbers A, B, and C are called the* **direction numbers** *of a vector or line if there is a scalar k such that kA, kB, and kC are the direction cosines of the vector or a vector collinear with the line.*

THEOREM 2.4.2   The components a_x, a_y, a_z of a vector $\mathbf{a}$ form a set of direction numbers for the vector $\mathbf{a}$.

THEOREM 2.4.3 $[\mathbf{a}\|\mathbf{b}] \Leftrightarrow \left[\begin{array}{l}\text{The direction numbers of } \mathbf{a} \text{ are} \\ \text{proportional to the direction} \\ \text{numbers of } \mathbf{b}.\end{array}\right.$

We leave the proofs of the preceding theorems as exercises.

SUMMARY

2.3 Base Vectors and Cartesian Coordinate Systems

DEFINITION 2.3.1 $[\mathbf{a}$ *is a unit vector*$] \Leftrightarrow [|\mathbf{a}| = 1]$

DEFINITION 2.3.2 *If every vector in a given space can be represented uniquely as a linear combination of a given set S of vectors in that space, then S is called a **basis** of the set of all vectors in that space and the elements of S are called **base vectors**.*

DEFINITION 2.3.3 Given the four points $O(0, 0, 0)$, $P_1(1, 0, 0)$, $P_2(0, 1, 0)$, $P_3(0, 0, 1)$ in a Cartesian coordinate space, we define:
$\mathbf{i} = \overrightarrow{OP_1}, \mathbf{j} = \overrightarrow{OP_2}, \mathbf{k} = \overrightarrow{OP_3}$.

THEOREM 2.3.1 (i) $\mathbf{i} \cdot \mathbf{i} = \mathbf{j} \cdot \mathbf{j} = \mathbf{k} \cdot \mathbf{k} = 1$
(ii) $\mathbf{i} \cdot \mathbf{j} = \mathbf{j} \cdot \mathbf{k} = \mathbf{k} \cdot \mathbf{i} = 0$

COROLLARY 2.3.1 The vectors $\mathbf{i}$, $\mathbf{j}$, and $\mathbf{k}$ form a basis for all vectors in 3-space.

THEOREM 2.3.2 $a_x = \text{comp}_i\, \mathbf{a} = \mathbf{a} \cdot \mathbf{i}, a_y = \text{comp}_j\, \mathbf{a} = \mathbf{a} \cdot \mathbf{j}, a_z = \text{comp}_k\, \mathbf{a} = \mathbf{a} \cdot \mathbf{k}$.

THEOREM 2.3.3 $\mathbf{a} + \mathbf{b} = (a_x + b_x)\mathbf{i} + (a_y + b_y)\mathbf{j} + (a_z + b_z)\mathbf{k}$.

THEOREM 2.3.4 $\mathbf{a} - \mathbf{b} = (a_x - b_x)\mathbf{i} + (a_y - b_y)\mathbf{j} + (a_z - b_z)\mathbf{k}$.

THEOREM 2.3.5 $h\mathbf{a} = (ha_x)\mathbf{i} + (ha_y)\mathbf{j} + (ha_z)\mathbf{k}$.

THEOREM 2.3.6 $[\mathbf{a} = \mathbf{0}] \Leftrightarrow [a_x = 0, a_y = 0, a_z = 0]$.

THEOREM 2.3.7 $[\mathbf{a} = \mathbf{b}] \Leftrightarrow [a_x = b_x, a_y = b_y, a_z = b_z]$.

THEOREM 2.3.8 $\mathbf{a} \cdot \mathbf{b} = a_x b_x + a_y b_y + a_z b_z$.

THEOREM 2.3.9 $\mathbf{a} \cdot \mathbf{a} = a_x^2 + a_y^2 + a_z^2 = a^2 = |\mathbf{a}|^2$.

THEOREM 2.3.10 Given the points $P_1(x_1, y_1, z_1)$ and $P_2(x_2, y_2, z_2)$, we have

$$\overrightarrow{P_1P_2} = (x_2 - x_1)\mathbf{i} + (y_2 - y_1)\mathbf{j} + (z_2 - z_1)\mathbf{k}.$$

THEOREM 2.3.11 If $\theta = \measuredangle(\mathbf{a}, \mathbf{b})$, then

$$\cos \theta = \frac{\mathbf{a} \cdot \mathbf{b}}{|\mathbf{a}|\,|\mathbf{b}|} = \frac{a_x b_x + a_y b_y + a_z b_z}{\sqrt{a_x^2 + a_y^2 + a_z^2}\,\sqrt{b_x^2 + b_y^2 + b_z^2}}$$

THEOREM 2.3.12   If **a** is a nonzero vector, then

$$\frac{\mathbf{a}}{|\mathbf{a}|} = \frac{a_x}{|\mathbf{a}|}\,\mathbf{i} + \frac{a_y}{|\mathbf{a}|}\,\mathbf{j} + \frac{a_z}{|\mathbf{a}|}\,\mathbf{k}$$

is a unit vector pointing in the direction of **a**.

2.4   Direction Cosines and Direction Numbers

DEFINITION 2.4.1   The **direction angles** of a vector **a** are the angles $\alpha = \sphericalangle(\mathbf{a}, \mathbf{i})$, $\beta = \sphericalangle(\mathbf{a}, \mathbf{j})$, $\gamma = \sphericalangle(\mathbf{a}, \mathbf{k})$.   The cosines of these angles are called the **direction cosines** of the vector **a**.

THEOREM 2.4.1   The direction cosines of a vector **a** are

$$\cos\alpha = \frac{a_x}{|\mathbf{a}|}, \cos\beta = \frac{a_y}{|\mathbf{a}|}, \cos\gamma = \frac{a_z}{|\mathbf{a}|}.$$

COROLLARY 2.4.1   The components of a unit vector having the same direction as **a** are the direction cosines of **a**.

COROLLARY 2.4.2   If $\cos\alpha$, $\cos\beta$, $\cos\gamma$ are the direction cosines of a vector **a**, then

$$(\cos\alpha)^2 + (\cos\beta)^2 + (\cos\gamma)^2 = 1.$$

DEFINITION 2.4.2   A set of numbers A, B, and C are called the **direction numbers** of a vector or line if there is a scalar k such that kA, kB, and kC are the direction cosines of the vector or a vector collinear with the line.

THEOREM 2.4.2   The components a_x, a_y, a_z of a vector **a** form a set of direction numbers for the vector **a**.

THEOREM 2.4.3 $[\mathbf{a}\|\mathbf{b}] \Leftrightarrow \left[\begin{array}{l} \text{The direction numbers of } \mathbf{a} \text{ are} \\ \text{proportional to the direction} \\ \text{numbers of } \mathbf{b}. \end{array}\right]$

PROBLEM SET #2.2

A. General

1. Prove Theorem 2.3.3, using Algebraic Laws 1.2.1–1.2.9 and appropriate definitions.

2. Prove Theorem 2.3.5, using Algebraic Laws 1.2.1–1.2.9 and appropriate definitions.

3. Prove Theorem 2.3.8.

4. Prove Theorem 2.3.10 and use the results to find the distance between $P(1, -2, 3)$ and $Q(-3, -1, 1)$.

5. Prove Theorem 2.3.11.

6. Prove Theorem 2.3.12.

7. Prove Corollaries 2.4.1 and 2.4.2.

8. Prove Theorem 2.4.2.

9. Prove Theorem 2.4.3.

10. Show that $\operatorname{comp}_a \mathbf{c} = \dfrac{\mathbf{a} \cdot \mathbf{c}}{|\mathbf{a}|}$, $\mathbf{a} \neq \mathbf{0}$.

11. Given $\mathbf{a} = 3\mathbf{i} + 2\mathbf{j} - \mathbf{k}$, $\mathbf{b} = \mathbf{i} - \mathbf{j} + 3\mathbf{k}$, $\mathbf{c} = -2\mathbf{i} - 3\mathbf{j} + \mathbf{k}$, find
   (a) $|\mathbf{a}|$ (b) $\mathbf{a} + \mathbf{b} - \mathbf{c}$ (c) $2\mathbf{b} - \mathbf{c}$ (d) $\mathbf{b} \cdot \mathbf{c}$
   (e) $\operatorname{comp}_a \mathbf{c}$ (f) $\mathbf{b} \cdot \mathbf{k}$ (g) $(\mathbf{a} \cdot \mathbf{b})\mathbf{c}$ (h) $\cos \measuredangle(\mathbf{a}, \mathbf{c})$

   (i) $\dfrac{\mathbf{a}}{|\mathbf{a}|}$ (j) the direction numbers and direction cosines of $\mathbf{a}$.

12. Given $\mathbf{a} = \mathbf{i} + 2\mathbf{j} - \mathbf{k}$, $\mathbf{b} = -\mathbf{i} - \mathbf{j} + \mathbf{k}$, $\mathbf{c} = 2\mathbf{i} - 3\mathbf{k}$, find the quantities (a) through (j) of Problem 11 above.

13. Given $P(3, 2, 1)$, $A(1, 1, 1)$, $B(0, 1, -2)$, find
   (a) $\overrightarrow{AB}$ (b) $|\overrightarrow{AP} - \overrightarrow{BP}|$ (c) $\overrightarrow{AP} \cdot \overrightarrow{BP}$ (d) $\cos \measuredangle(\overrightarrow{PA}, \overrightarrow{PB})$

   (e) the direction numbers and direction cosines of $\overrightarrow{AP}$.

14. Given $P(2, 0, -1)$, $A(1, 1, -1)$, $B(-1, 2, 0)$, find the quantities (a) through (e) of Problem 13 above.

15. Find a vector that is parallel to the line joining $P(2, -3, 5)$ and $Q(-1, -3, 1)$.

16. Find the direction cosines of a directed line through the origin having the same direction as the vector $\overrightarrow{PQ}$ where the coordinates of P and Q are $(1, -2, 3)$ and $(2, 3, -1)$, respectively. How many sets of direction cosines does a nondirected line have?

17. If $\mathbf{a} = 3\mathbf{i} - 2\mathbf{j}$ and $\mathbf{b} = \mathbf{i} + 2\mathbf{j}$ and $\mathbf{c} = 2\mathbf{i} - 3\mathbf{j}$, then find scalars α and β so that $\mathbf{c} = \alpha\mathbf{a} + \beta\mathbf{b}$. What would one have to show to conclude that $\mathbf{a}$ and $\mathbf{b}$ form a basis for all vectors in the xy-plane? Do the vectors $\mathbf{r} = \frac{1}{2}\mathbf{i} - 3\mathbf{k}$ and $\mathbf{s} = -\mathbf{i} + 6\mathbf{k}$ form a basis in the xy-plane? Explain.

18. If $\mathbf{a} = \mathbf{i} + 3\mathbf{j} - 2\mathbf{k}$, $\mathbf{b} = 2\mathbf{i} - \mathbf{j} + \mathbf{k}$, $\mathbf{c} = -2\mathbf{i} + \mathbf{j} - 3\mathbf{k}$, and $\mathbf{d} = 3\mathbf{i} + 2\mathbf{j} + 5\mathbf{k}$, then find scalars α, β, γ such that $\mathbf{d} = \alpha\mathbf{a} + \beta\mathbf{b} + \gamma\mathbf{c}$. What would one have to show to conclude that $\mathbf{a}$, $\mathbf{b}$, and $\mathbf{c}$ form a basis in 3-space?

B. Geometry

ON THE DISTANCE FROM A POINT TO A LINE IN THE PLANE

1. (a) Show that the vector $\mathbf{N} = \alpha\mathbf{i} + \beta\mathbf{j}$ is perpendiculr to the line $\alpha x + \beta y + \gamma = 0$ in the xy-plane.

(b) Find a unit vector perpendicular to the line $2x - 3y + 2 = 0$ in the xy-plane.

(c) Show that the distance from the point $P(x_1, y_1)$ to the line $ax + by + c = 0$ is given by

$$d = \left| \frac{\mathbf{N} \cdot \vec{QP}}{|\mathbf{N}|} \right|$$

where $\mathbf{N}$ is normal to $ax + by + c = 0$ and Q is a point on the line.

(d) Show that a nonvector form for d is:

$$d = \frac{|ax_1 + by_1 + c|}{\sqrt{a^2 + b^2}}$$

(e) Find the distance from $(3, 5)$ to $2x - 3y + 5 = 0$.

2. How much closer is the point $P(2, 3)$ to the line $2y = x$ than to the line $2y = -x - 1$?

ON PLANES

3. (a) Write the equation of the plane through $P(x_0, y_0, z_0)$ perpendicular to the vector $\mathbf{N} = A\mathbf{i} + B\mathbf{j} + C\mathbf{k}$
  (i) in vector form.
  (ii) in Cartesian coordinate form.
  (*Hint:* Let $Q(x, y, z)$ be an arbitrary point in the plane; then $\vec{PQ} \cdot \mathbf{N} = ?$)

(b) Find the equation of the plane that is perpendicular to the line through $A(1, -1, -1)$ and $B(0, 1, 1)$ and that passes through the point $P(2, 1, 5)$.

4. Find the equation of the plane that passes through $Q(1, -1, 3)$ and has the vector $\mathbf{N} = 2\mathbf{i} + \mathbf{j} - \mathbf{k}$ as a normal.

5. Given the plane $Ax + By + Cz = D$, show that A, B, and C are direction numbers of a line perpendicular to the plane. Find a unit vector normal to the plane $3x - 2y + 5z = -6$.

6. Find the angle between the planes $3x + y - z + 4 = 0$ and $2x - 3y - 8z + 1 = 0$. (*Hint:* Notice the relationship between the dihedral angle and the angle between the normals to the planes.)

7. Find a formula for the distance from a point $P(x_1, y_1, z_1)$ to a plane $Ax + By + Cz + D = 0$
  (a) in vector form. (*Hint:* see Problem 1 above.)
  (b) in nonvector, Cartesian coordinate form.

8. (a) Find the distance from the origin to the plane $2x + y - 3z + 14 = 0$.
  (b) Find the distance from $(3, 2, -1)$ to the plane $2x + y - 3z + 14 = 0$.

ON THE SPHERE

9. (a) Find the equation of a sphere with center at $P(x_0, y_0, z_0)$ and radius r units
  (i) in vector form. (ii) in Cartesian coordinates.
 (b) What is the equation of the sphere with center at $(-2, -1, 5)$ and radius equal to 8 units?

10. Find the equation of the plane that is tangent to the unit sphere, with center at the origin, and parallel to the plane $3x - 2y + 5z - 6 = 0$.

MISCELLANEOUS PROBLEMS

11. Given $\mathbf{a} = \mathbf{i} - 3\mathbf{j} + 5\mathbf{k}$, $\mathbf{b} = 2\mathbf{i} + \mathbf{j} - 4\mathbf{k}$, and $\mathbf{c} = 3\mathbf{i} - 2\mathbf{j} + \mathbf{k}$, show that the vectors can be placed to form a right triangle.

12. Let $\mathbf{a} = (\cos \alpha_1)\mathbf{i} + (\cos \beta_1)\mathbf{j}$ and $\mathbf{b} = (\cos \alpha_2)\mathbf{i} + (\cos \beta_2)\mathbf{j}$ be two unit vectors in the xy-coordinate plane. Prove that
  (i) $\cos (\alpha_1 + \alpha_2) = \cos \alpha_1 \cos \alpha_2 - \sin \alpha_1 \sin \alpha_2$
  (ii) $\cos (\alpha_1 - \alpha_2) = \cos \alpha_1 \cos \alpha_2 + \sin \alpha_1 \sin \alpha_2$

13. If O is the origin, find the locus of all points P such that

$$|\overrightarrow{OP}| = 5 \quad \text{and} \quad \frac{\overrightarrow{OP} \cdot \mathbf{j}}{|\overrightarrow{OP}|} = \frac{1}{2}.$$

14. Find the angle between the diagonal of a cube and one of its edges.

15. Find a vector perpendicular to the vectors $\mathbf{a} = \mathbf{i} - \mathbf{j} + \mathbf{k}$ and $\mathbf{b} = 2\mathbf{i} + 3\mathbf{j} - \mathbf{k}$.

16. Given vectors $\mathbf{a} = 2\mathbf{i} + \mathbf{j} - \mathbf{k}$ and $\mathbf{b} = -\mathbf{i} + 3\mathbf{j} - \mathbf{k}$ with their initial ends at the origin, show that the vector joining the terminal points of $\mathbf{a}$ and $\mathbf{b}$ is parallel to the xy-coordinate plane. Find its magnitude.

C. Physics

1. Find the work done by a constant force $\mathbf{F} = 2\mathbf{i} - 3\mathbf{j} + \mathbf{k}$ on an object that moves along the straight line segments connecting $P_1(3, -1, 2)$, $P_2(1, 2, 0)$, $P_3(-5, 2, -3)$, and $P_4(4, 1, -2)$.

2. Find the work done by the force in Problem 1 if the path is the single line segment joining P_1 and P_4.

3. What is the physical significance of the general distributive law

$$\mathbf{F} \cdot (\mathbf{a}_1 + \mathbf{a}_2 + \cdots + \mathbf{a}_n) = \mathbf{F} \cdot \mathbf{a}_1 + \mathbf{F} \cdot \mathbf{a}_2 + \cdots + \mathbf{F} \cdot \mathbf{a}_n?$$

How are Problems 1 and 2 related to this law?

4. If the vector $\mathbf{V}$ represents the magnitude and direction of the velocity of a fluid flowing through a plane surface area, and if $\mathbf{N}$ is a unit normal to the surface, then show that

$$\mathbf{V} \cdot \mathbf{N} = \begin{bmatrix} \text{The volume of fluid passing through} \\ \text{a unit of area per unit time.} \end{bmatrix}$$

(Assume that the flow lines of the fluid are parallel.)

3

the vector product
triple products
vector identities

3.1 THE VECTOR (CROSS) PRODUCT

In preceding sections we defined and discussed two types of products that involve vectors: the product of a scalar and a vector and the dot (or scalar) product of two vectors.   A third type of vector product that has many applications in physics (e.g., mechanics, electricity, and fluid dynamics) and geometry is now defined.

DEFINITION 3.1.1 *The **vector** (or **cross**) **product** of two vectors* **a** *and* **b** *is a third vector* **c**, *denoted by* **a** **×** **b**, *such that*

(i) $|\mathbf{c}| = |\mathbf{a}|\,|\mathbf{b}|\sin\theta$ *where* $\theta = \measuredangle(\mathbf{a}, \mathbf{b})$

(ii) $\mathbf{c} \cdot \mathbf{a} = \mathbf{c} \cdot \mathbf{b} = 0$ *(i.e.,* $\mathbf{c} \perp \mathbf{a}$ *and* $\mathbf{c} \perp \mathbf{b}$ *)*

(iii) *the sense of* **c** *is chosen so that* **a**, **b**, *and* **c** *form a right-hand system.**

---

* When **a** is rotated through an angle less than 180° into **b**, the forward motion of a right-hand screw will be in the positive **c**-direction.

44

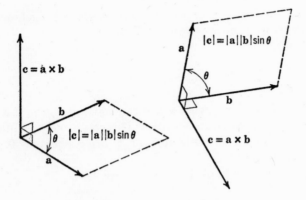

FIGURE 3.1.1   The vector product.

If the vectors **a**, **b**, and **c** are placed so that their initial ends are at a common point, **a** and **b** can be interpreted as adjacent sides of a parallelogram with **c** a normal to the parallelogram. The vector **c** will have a magnitude equal to the area of the parallelogram.

In many applications the following three properties of the vector product are found useful.   The reader should find them easy to prove.

THEOREM 3.1.1 $|a \times b|$ is the area of a parallelogram with **a** and **b** as adjacent sides.

THEOREM 3.1.2 $[a \times b = 0] \Leftrightarrow [a$ and **b** are collinear].

THEOREM 3.1.3   The vector $\dfrac{a \times b}{|a \times b|}$ is a unit normal vector to any plane parallel to both **a** and **b**.

The vector product has the following algebraic properties:

ALGEBRAIC LAWS

3.1.1	$a \times b = -(b \times a)$	anticommutative law
3.1.2	$a \times (b + c) = (a \times b) + (a \times c)$	distributive law
3.1.3	$a \times (hb) = (ha) \times b = h(a \times b)$	associative law for h a scalar

A few remarks are in order regarding several algebraic laws that the vector and scalar products do **not** obey.

(a) In the algebra of real numbers we have

$$[ab = 0] \Leftrightarrow [a = 0 \text{ or } b = 0].$$

However, in vector algebra $\mathbf{a} \times \mathbf{b}$ can equal zero even if $\mathbf{a}$ and $\mathbf{b}$ are both nonzero vectors. (See Theorem 3.1.2.) Note that $\mathbf{a} \cdot \mathbf{b}$ can equal zero without $\mathbf{a}$ or $\mathbf{b}$ being equal to $\mathbf{0}$. (See Theorem 2.1.3.)

(b) In the algebra of real numbers we have

$$a(bc) = (ab)c.$$

That is, there is an associative law for combinations of successive products. Try to show geometrically why $\mathbf{a} \times (\mathbf{b} \times \mathbf{c}) \neq (\mathbf{a} \times \mathbf{b}) \times \mathbf{c}$. (*Hint:* $\mathbf{a} \times (\mathbf{b} \times \mathbf{c})$ is a vector in a plane parallel to $\mathbf{b}$ and $\mathbf{c}$.) It should be evident that the associative law for scalar products, $\mathbf{a} \cdot (\mathbf{b} \cdot \mathbf{c}) = (\mathbf{a} \cdot \mathbf{b}) \cdot \mathbf{c}$, is undefined.

(c) In the algebra of real numbers we have

$$[ab = ac, a \neq 0] \Leftrightarrow [b = c].$$

That is, there is a cancellation law. This law does not hold for either the scalar or vector products. (See Theorems 2.1.3 and 3.1.2.)

3.2 THE VECTOR PRODUCT IN COMPONENT FORM

It is convenient to note the following special vector products:

THEOREM 3.2.1 $\mathbf{a} \times \mathbf{a} = \mathbf{0}$

THEOREM 3.2.2 $\mathbf{i} \times \mathbf{i} = \mathbf{0}, \quad \mathbf{i} \times \mathbf{j} = \mathbf{k}, \quad \mathbf{i} \times \mathbf{k} = -\mathbf{j}$
$\mathbf{j} \times \mathbf{i} = -\mathbf{k}, \quad \mathbf{j} \times \mathbf{j} = \mathbf{0}, \quad \mathbf{j} \times \mathbf{k} = \mathbf{i}$
$\mathbf{k} \times \mathbf{i} = \mathbf{j}, \quad \mathbf{k} \times \mathbf{j} = -\mathbf{i}, \quad \mathbf{k} \times \mathbf{k} = \mathbf{0}$

The vector product, $\mathbf{a} \times \mathbf{b}$, can be expressed in terms of the components of $\mathbf{a}$ and $\mathbf{b}$ as shown below. The component form of the vector product is extremely important in applications.

THEOREM 3.2.3   Let $\mathbf{a} = a_x\mathbf{i} + a_y\mathbf{j} + a_z\mathbf{k}$ and $\mathbf{b} = b_x\mathbf{i} + b_y\mathbf{j} + b_z\mathbf{k}$; then

$$\mathbf{a} \times \mathbf{b} = (a_yb_z - a_zb_y)\mathbf{i} + (a_zb_x - a_xb_z)\mathbf{j} + (a_xb_y - a_yb_x)\mathbf{k}$$

or, we can write

$$\mathbf{a} \times \mathbf{b} = \begin{vmatrix} \mathbf{i} & \mathbf{j} & \mathbf{k} \\ a_x & a_y & a_z \\ b_x & b_y & b_z \end{vmatrix}^*$$

Proof $\mathbf{a} \times \mathbf{b} = (a_x\mathbf{i} + a_y\mathbf{j} + a_z\mathbf{k}) \times (b_x\mathbf{i} + b_y\mathbf{j} + b_z\mathbf{k})$

$\quad = (a_x\mathbf{i} + a_y\mathbf{j} + a_z\mathbf{k}) \times b_x\mathbf{i}$
$\quad + (a_x\mathbf{i} + a_y\mathbf{j} + a_z\mathbf{k}) \times b_y\mathbf{j}$
$\quad + (a_x\mathbf{i} + a_y\mathbf{j} + a_z\mathbf{k}) \times b_z\mathbf{k}$

$\mathbf{a} \times \mathbf{b} = a_xb_x(\mathbf{i} \times \mathbf{i}) + a_yb_x(\mathbf{j} \times \mathbf{i}) + a_zb_x(\mathbf{k} \times \mathbf{i})$
$\quad + a_xb_y(\mathbf{i} \times \mathbf{j}) + a_yb_y(\mathbf{j} \times \mathbf{j}) + a_zb_y(\mathbf{k} \times \mathbf{j})$
$\quad + a_xb_z(\mathbf{i} \times \mathbf{k}) + a_yb_z(\mathbf{j} \times \mathbf{k}) + a_zb_z(\mathbf{k} \times \mathbf{k})$

$\mathbf{a} \times \mathbf{b} = 0 + a_yb_x(-\mathbf{k}) + a_zb_x\mathbf{j} + a_xb_y\mathbf{k}$
$\quad + 0 + a_zb_y(-\mathbf{i}) + a_xb_z(-\mathbf{j}) + a_yb_z\mathbf{i} + 0$

$\mathbf{a} \times \mathbf{b} = (a_yb_z - a_zb_y)\mathbf{i} + (a_zb_x - a_xb_z)\mathbf{j} + (a_xb_y - a_yb_x)\mathbf{k}$

$$= \begin{vmatrix} a_y & a_z \\ b_y & b_z \end{vmatrix}\mathbf{i} + \begin{vmatrix} a_z & a_x \\ b_z & b_x \end{vmatrix}\mathbf{j} + \begin{vmatrix} a_x & a_y \\ b_x & b_y \end{vmatrix}\mathbf{k}$$

$$\mathbf{a} \times \mathbf{b} = \begin{vmatrix} \mathbf{i} & \mathbf{j} & \mathbf{k} \\ a_x & a_y & a_z \\ b_x & b_y & b_z \end{vmatrix}.$$

EXAMPLE 3.2.1   If $\mathbf{a} = 2\mathbf{i} - \mathbf{j} + \mathbf{k}$ and $\mathbf{b} = \mathbf{i} - 2\mathbf{k}$, find $\mathbf{a} \times \mathbf{b}$.

(a) Using Theorem 3.2.3, we have

$$\mathbf{a} \times \mathbf{b} = \begin{vmatrix} \mathbf{i} & \mathbf{j} & \mathbf{k} \\ 2 & -1 & 1 \\ 1 & 0 & -2 \end{vmatrix} = 2\mathbf{i} + 5\mathbf{j} + \mathbf{k}$$

(b) Using Algebraic Laws 3.1.1 through 3.1.3 and Theorem 3.2.2, we have

$\mathbf{a} \times \mathbf{b} = (2\mathbf{i} - \mathbf{j} + \mathbf{k}) \times (\mathbf{i} - 2\mathbf{k})$
$\quad = (2\mathbf{i} - \mathbf{j} + \mathbf{k}) \times \mathbf{i} + (2\mathbf{i} - \mathbf{j} + \mathbf{k}) \times (-2\mathbf{k})$
$\quad = 2(\mathbf{i} \times \mathbf{i}) - (\mathbf{j} \times \mathbf{i}) + (\mathbf{k} \times \mathbf{i})$
$\quad - 4(\mathbf{i} \times \mathbf{k}) + 2(\mathbf{j} \times \mathbf{k}) - 2(\mathbf{k} \times \mathbf{k})$
$\quad = \mathbf{k} + \mathbf{j} + 4\mathbf{j} + 2\mathbf{i}$
$\quad = 2\mathbf{i} + 5\mathbf{j} + \mathbf{k}$

* This determinant form is to be expanded by minors, using the top row.

SUMMARY

3.1 The Vector (Cross) Product

DEFINITION 3.1.1 *The* **vector** (cross) **product** *of two vectors* **a** *and* **b** *is a third vector* **c**, *denoted by* **a** × **b**, *such that*

(i) $|\mathbf{c}| = |\mathbf{a}| \, |\mathbf{b}| \sin \theta$ *where* $\theta = \sphericalangle(\mathbf{a}, \mathbf{b})$

(ii) $\mathbf{c} \cdot \mathbf{a} = \mathbf{c} \cdot \mathbf{b} = 0$ (*i.e.,* **c** ⊥ **a** *and* **c** ⊥ **b**)

(iii) *the sense of* **c** *is chosen so that* **a**, **b**, *and* **c** *form a right-hand system.*

THEOREM 3.1.1 $|\mathbf{a} \times \mathbf{b}|$ is the area of a parallelogram with **a** and **b** as adjacent sides.

THEOREM 3.1.2 $[\mathbf{a} \times \mathbf{b} = \mathbf{0}] \Leftrightarrow$ [**a** and **b** are collinear].

THEOREM 3.1.3 The vector $\dfrac{\mathbf{a} \times \mathbf{b}}{|\mathbf{a} \times \mathbf{b}|}$ is a unit normal vector to any plane parallel to both **a** and **b**.

ALGEBRAIC LAWS

3.1.1 $\mathbf{a} \times \mathbf{b} = -(\mathbf{b} \times \mathbf{a})$ anticommutative law

3.1.2 $\mathbf{a} \times (\mathbf{b} + \mathbf{c}) = (\mathbf{a} \times \mathbf{b}) + (\mathbf{a} \times \mathbf{c})$ distributive law

3.1.3 $\mathbf{a} \times (h\mathbf{b}) = (h\mathbf{a}) \times \mathbf{b} = h(\mathbf{a} \times \mathbf{b})$ associative law

                                                     for h a scalar

3.2 The Vector Product in Component Form

THEOREM 3.2.1 $\mathbf{a} \times \mathbf{a} = \mathbf{0}$

THEOREM 3.2.2 $\mathbf{i} \times \mathbf{i} = \mathbf{0}, \qquad \mathbf{i} \times \mathbf{j} = \mathbf{k}, \qquad \mathbf{i} \times \mathbf{k} = -\mathbf{j}$

                      $\mathbf{j} \times \mathbf{i} = -\mathbf{k}, \qquad \mathbf{j} \times \mathbf{j} = \mathbf{0}, \qquad \mathbf{j} \times \mathbf{k} = \mathbf{i}$

                      $\mathbf{k} \times \mathbf{i} = \mathbf{j}, \qquad \mathbf{k} \times \mathbf{j} = -\mathbf{i}, \qquad \mathbf{k} \times \mathbf{k} = \mathbf{0}$

THEOREM 3.2.3 Let $\mathbf{a} = a_x\mathbf{i} + a_y\mathbf{j} + a_z\mathbf{k}$ and $\mathbf{b} = b_x\mathbf{i} + b_y\mathbf{j} + b_z\mathbf{k}$; then

$$\mathbf{a} \times \mathbf{b} = (a_yb_z - a_zb_y)\mathbf{i} + (a_zb_x - a_xb_z)\mathbf{j} + (a_xb_y - a_yb_x)\mathbf{k}$$

or

$$\mathbf{a} \times \mathbf{b} = \begin{vmatrix} \mathbf{i} & \mathbf{j} & \mathbf{k} \\ a_x & a_y & a_z \\ b_x & b_y & b_z \end{vmatrix}.$$

PROBLEM SET #3.1

A. General

1. If $\mathbf{a} = \mathbf{i} + \mathbf{j} - 2\mathbf{k}$, $\mathbf{b} = 2\mathbf{i} - \mathbf{k}$, and $\mathbf{c} = -\mathbf{j} + 2\mathbf{k}$, then find
   (a) $\mathbf{a} \times \mathbf{b}$ (b) $\mathbf{a} \times \mathbf{c}$ (c) $\mathbf{b} \times \mathbf{a}$
   (d) $\mathbf{a} \times (\mathbf{b} + \mathbf{c})$ (e) $\mathbf{a} \times (\mathbf{b} \times \mathbf{c})$

2. If $\mathbf{a} = 2\mathbf{i} - \mathbf{j} - \mathbf{k}$, $\mathbf{b} = 2\mathbf{j} + \mathbf{k}$, and $\mathbf{c} = \mathbf{i} - \mathbf{k}$, then find the quantities (a) through (e) of Problem 1 above.

3. If $\mathbf{a} = 3\mathbf{i} + 2\mathbf{k}$ and $\mathbf{b} = \mathbf{j} - 3\mathbf{k}$, find $\mathbf{a} \times \mathbf{b}$ directly without using Theorem 3.2.3.

4. Without using Theorem 3.2.3, find $\mathbf{a} \times \mathbf{b}$, where $\mathbf{a} = \mathbf{i} - 3\mathbf{k}$ and $\mathbf{b} = \mathbf{j} + 2\mathbf{k}$.

5. Given $\mathbf{a} = 2\mathbf{i} + \mathbf{j} - \mathbf{k}$ and $\mathbf{b} = -3\mathbf{i} + 2\mathbf{j} + \mathbf{k}$, find
   (a) the unit vector perpendicular to both $\mathbf{a}$ and $\mathbf{b}$.
   (b) the sine of the angle between $\mathbf{a}$ and $\mathbf{b}$ when taken from a common initial point.

6. Given $\mathbf{a} = \mathbf{i} - 2\mathbf{k}$ and $\mathbf{b} = 2\mathbf{i} + 3\mathbf{j}$, find the quantities (a) and (b) of Prob. 5 above.

7. Prove that $(\mathbf{a} + \mathbf{b}) \times (\mathbf{a} - \mathbf{b}) = -2\mathbf{a} \times \mathbf{b}$.

8. If $\mathbf{a} + \mathbf{b} + \mathbf{c} = \mathbf{0}$, show that $\mathbf{a} \times \mathbf{b} = \mathbf{b} \times \mathbf{c} = \mathbf{c} \times \mathbf{a}$.

9. Show that $\mathbf{a} \times \mathbf{b}$ can equal $\mathbf{a} \times \mathbf{c}$ without $\mathbf{b}$ being equal to $\mathbf{c}$, that is, that the "cancellation law" does not hold for the vector product.

10. Show that if $\mathbf{b} = \mathbf{c} + \alpha\mathbf{a}$, then $\mathbf{a} \times \mathbf{b} = \mathbf{a} \times \mathbf{c}$.

11. Given that $\mathbf{a} \neq \mathbf{0}$ and both $\mathbf{a} \cdot \mathbf{b} = \mathbf{a} \cdot \mathbf{c}$ and $\mathbf{a} \times \mathbf{b} = \mathbf{a} \times \mathbf{c}$, show that $\mathbf{b} = \mathbf{c}$.

12. Prove that, in general, $(\mathbf{a} \times \mathbf{b}) \times \mathbf{c} \neq \mathbf{a} \times (\mathbf{b} \times \mathbf{c})$.
(*Hint:* $(\mathbf{a} \times \mathbf{b}) \times \mathbf{c}$ is a vector lying in the plane parallel to the vectors $\mathbf{a}$ and $\mathbf{b}$.)

13. Prove that $(\mathbf{a} \times \mathbf{b}) \cdot (\mathbf{c} \times \mathbf{d}) = (\mathbf{a} \cdot \mathbf{c})(\mathbf{b} \cdot \mathbf{d}) - (\mathbf{a} \cdot \mathbf{d})(\mathbf{b} \cdot \mathbf{c})$.

14. Prove that

$$(\mathbf{a} \times \mathbf{b}) \cdot (\mathbf{a} \times \mathbf{b}) = \begin{vmatrix} \mathbf{a} \cdot \mathbf{a} & \mathbf{a} \cdot \mathbf{b} \\ \mathbf{a} \cdot \mathbf{b} & \mathbf{b} \cdot \mathbf{b} \end{vmatrix}$$

B. Geometric

ON EQUATIONS OF LINES IN 3-SPACE

1. Prove that $[\mathbf{a} \times \mathbf{b} = \mathbf{0}] \Leftrightarrow [\mathbf{a}$ and $\mathbf{b}$ are collinear$]$.

2. Which of the following three vectors are collinear:

$$\mathbf{a} = -\mathbf{i} - \tfrac{1}{2}\mathbf{j} + 3\mathbf{k}, \quad \mathbf{b} = -\mathbf{i} + \tfrac{1}{2}\mathbf{j} - 3\mathbf{k}, \quad \mathbf{c} = 2\mathbf{i} - \mathbf{j} + 6\mathbf{k}?$$

3. (a) Show that $\overrightarrow{P_1P_2} \times \overrightarrow{P_1P} = \mathbf{0}$ is the vector equation of a line through the fixed points P_1 and P_2, where P is a variable point. (*Hint:* Use the results of Problem 1.)
   (b) Assign coordinates to P_1, P_2, and P in part (a) and obtain the equation of the line through $P_1(x_1, y_1, z_1)$ and $P_2(x_2, y_2, z_2)$ in the nonvector form

$$\frac{x - x_1}{x_2 - x_1} = \frac{y - y_1}{y_2 - y_1} = \frac{z - z_1}{z_2 - z_1}$$

called **the symmetric form** of the equation of a line.

(c) If one of the denominators in (b) is equal to zero, say, $y_1 = y_2$, how is the symmetric form modified?

(d) If two of the denominators are equal to zero, say, $y_1 = y_2$ and $z_1 = z_2$, how is the symmetric form modified?

4. (a) Write the equation of the line through $P_1(2, -1, 3)$ and $P_2(1, -2, -2)$ in the symmetric form.

(b) Write the equation of the line through $P_1(5, -7, 1)$ and $P_2(-4, -7, 3)$.

(c) Write the equation of the line through $P_1(-1, 2, 3)$ and $P_2(-1, 4, 3)$.

5. (a) Show that $\overrightarrow{P_1P} \times \mathbf{a} = \mathbf{0}$ is the vector equation of a line that passes through P_1 and has direction numbers a_x, a_y, a_z.

(b) From (a) conclude that

$$\frac{x - x_1}{a_x} = \frac{y - y_1}{a_y} = \frac{z - z_1}{a_z}$$

is the equation of a line that passes through (x_1, y_1, z_1), which has direction numbers $a_x, a_y,$ and a_z. Note that the symmetric form (Problem 3) is a special case of this form such that $(x_2 - x_1)$, $(y_2 - y_1)$, and $(z_2 - z_1)$ are direction numbers.

(c) What effect do zero direction numbers have on the equation of part (b) above?

6. (a) Write the equation of the line through the point $(2, -1, 0)$ parallel to the line through $(3, -1, -2)$ and $(-1, 2, 1)$.

(b) Write the equation of the line through the point $(3, -2, 1)$ parallel to the vector $2\mathbf{i} - 3\mathbf{j}$.

(c) Write the equation of the line through the point $(3, -2, 1)$ parallel to the vector $-4\mathbf{k}$.

7. (a) Show that $\overrightarrow{P_1P} = t\mathbf{a}, -\infty < t < \infty$, is another vector equation of the line that passes through the point P_1 and has direction numbers a_x, a_y, a_z.

(b) Deduce from part (a) that

$$\left. \begin{array}{l} x = x_1 + a_x t \\ y = y_1 + a_y t \\ z = z_1 + a_z t \end{array} \right\} \text{ for } -\infty < t < \infty$$

is the corresponding nonvector form of the equation of the same line. This form is called **the parametric form** of the equation of a line.

(c) Transform the parametric form to the symmetric form of the equation of a line.

8. Write the parametric form of equations of lines with the following properties:
   (a) through $(2, -1, 3)$ and $(1, -2, -2)$.
   (b) through $(5, -7, 1)$ and $(-4, -7, 3)$.
   (c) through $(-1, 2, 3)$ and $(-1, 4, 3)$.
   (d) through $(2, -1, 0)$ parallel to the line through $(3, -1, -2)$ and $(-1, 2, 1)$.
   (e) through $(3, -2, 1)$ parallel to the vector $2\mathbf{i} - 3\mathbf{j}$.
   (f) through $(3, -2, 1)$ parallel to the vector $-4\mathbf{k}$.

ON EQUATIONS OF A PLANE IN 3-SPACE

Recall the following results from Problem Set #2.2, Problems 3, 5, 7 on geometry.

(i) $\overrightarrow{P_1P} \cdot \mathbf{N} = 0$ is the vector equation of a plane through P_1 with normal $\mathbf{N}$.

(ii) $A(x - x_1) + B(y - y_1) + C(z - z_1) = 0$ or $Ax + By + Cz = D$ are the equations of a plane in Cartesian coordinates.

(iii) A, B, and C are the direction numbers of a line normal to the plane $Ax + By + Cz = D$.

(iv) The distance from a point P_1 to a plane $Ax + By + Cz = D$ is given by the formula

$$d = \left| \frac{\mathbf{N}}{|\mathbf{N}|} \cdot \overrightarrow{P_0P_1} \right| = \frac{|Ax_1 + By_1 + Cz_1 - D|}{\sqrt{A^2 + B^2 + C^2}}$$

where $\mathbf{N} = A\mathbf{i} + B\mathbf{j} + C\mathbf{k}$ and $Ax_0 + By_0 + Cz_0 = D$.

9. (a) Show that the vector equation of a plane through three fixed points P_1, P_2, P_3, which are noncollinear, is

$$(\overrightarrow{P_1P_2} \times \overrightarrow{P_1P_3}) \cdot \overrightarrow{P_1P} = 0.$$

(b) Show that the vector form of the equation in part (a) is equivalent to the nonvector form

$$\begin{vmatrix} (x - x_1) & (y - y_1) & (z - z_1) \\ (x_2 - x_1) & (y_2 - y_1) & (z_2 - z_1) \\ (x_3 - x_1) & (y_3 - y_1) & (z_3 - z_1) \end{vmatrix} = 0.$$

10. Find a vector $\mathbf{N}$ normal to the plane containing the three points $P_1(0, -1, 2)$, $P_2(1, 2, -3)$, and $P_3(-1, 2, 3)$.

11. Find the nonvector equation of the plane containing the three points of Problem 10 above.

12. (a) Find a unit vector $\mathbf{N}$ normal to the two lines

$$\frac{x - 2}{3} = \frac{y - 1}{2} = \frac{z + 3}{-1} \quad \text{and} \quad \frac{x - 4}{2} = \frac{y}{-3} = \frac{z - 4}{2}.$$

(b) Find the equation of a plane parallel to the lines

$$\frac{x-2}{3} = \frac{y-1}{2} = \frac{z+3}{-1} \quad \text{and} \quad \frac{x-4}{2} = \frac{y}{-3} = \frac{z-4}{2}.$$

13. (a) Find the equations of the parallel planes which contain the lines

$$\frac{x-2}{3} = \frac{y-1}{2} = \frac{z+3}{-1} \quad \text{and} \quad \frac{x-4}{2} = \frac{y}{-3} = \frac{z-4}{2}$$

respectively.

(b) Find the distance between the two planes.

14. Let m_1 and m_2 denote the two lines in the preceding problem. What is the geometric significance of

$$\left(\frac{\mathbf{a} \times \mathbf{b}}{|\mathbf{a} \times \mathbf{b}|} \right) \cdot (\overrightarrow{P_0 P_1})$$

where $\mathbf{a}$ is a vector collinear with m_1,

$\mathbf{b}$ is a vector collinear with m_2,

P_0 is a point on m_1,

P_1 is a point on m_2?

How is this question related to Problem 13?

ON THE DISTANCE FROM A POINT TO A PLANE

15. Show that the distance from a point P_0, noncoplanar with the points P_1, P_2, and P_3, to the plane determined by P_1, P_2, and P_3 is given by

$$d = \left| \frac{\overrightarrow{P_1 P_2} \times \overrightarrow{P_1 P_3}}{|\overrightarrow{P_1 P_2} \times \overrightarrow{P_1 P_3}|} \cdot \overrightarrow{P_1 P_0} \right|.$$

16. Find the distance from the point $(2, 4, 6)$ to the plane determined by the three points $(0, -1, 2)$, $(1, 2, -3)$, and $(-1, 2, 3)$.

ON THE DISTANCE BETWEEN TWO LINES

17. Given four noncoplanar points A, B, C, and D, show that the distance between the two lines determined by A, B and C, D, respectively, is given by

$$d = \left| \frac{\overrightarrow{AB} \times \overrightarrow{CD}}{|\overrightarrow{AB} \times \overrightarrow{CD}|} \cdot \overrightarrow{AC} \right|.$$

18. Given the points $A(-1, -2, 2)$, $B(-2, 1, -1)$, $C(2, -1, 3)$, and $D(3, 0, 3)$, find the distance between the lines determined by A, B and C, D, respectively.

19. Find the distance between the two lines

$$\frac{x-2}{3} = \frac{y-1}{2} = \frac{z+3}{-1} \quad \text{and} \quad \frac{x-4}{2} = \frac{y}{-3} = \frac{z-4}{2}.$$

MISCELLANEOUS PROBLEMS IN GEOMETRY

20. Given points $A(1, 1, 1)$, $B(2, 3, 0)$, $C(0, 5, 1)$, and $D(1, 2, 4)$, find
    (a) the area of triangle ABC.
    (b) the distance from D to the plane containing the triangle ABC.
    (c) the distance between the lines determined by BC and AD.
        (*Hint:* the distance is measured along a segment which is
        mutually perpendicular to the two lines.)

21. If $A(x_1, y_1, z_1)$, $B(x_2, y_2, z_2)$, and $C(x_3, y_3, z_3)$ are the vertices of a
triangle, find a formula in Cartesian coordinates for the area of the
triangle ABC.

22. Find the area of a triangle with vertices $A(1, -1, 2)$, $B(-2, 1, 1)$,
$C(3, 2, 1)$.

23. If **a**, **b**, **c** with a common initial point O represent three adjacent
edges of a parallelepiped, find a formula for the volume of the parallelepiped in vector and scalar forms.

24. Find the volume of the tetrahedron with vertices $A(1, 2, 1)$, $B(3,$
$5, 0)$, $C(0, 3, -1)$, and $D(2, 4, 4)$.

25. Use the vector product in proving the law of sines for plane triangles:

$$\frac{a}{\sin A} = \frac{b}{\sin B} = \frac{c}{\sin C}.$$

26. Let A_1, A_2, A_3, and A_4 be the areas of the four faces of a tetra-
hedron. Let $\mathbf{n}_1$, $\mathbf{n}_2$, $\mathbf{n}_3$, and $\mathbf{n}_4$ be outward normals to the respective
faces with magnitudes equal to the corresponding areas. Show that
$\mathbf{n}_1 + \mathbf{n}_2 + \mathbf{n}_3 + \mathbf{n}_4 = \mathbf{0}$.

C. Physics

1. The law of refraction of light, known as Snell's Law, states: "The
ratio of the sine of the angle of incidence to the sine of the angle of refraction is equal to the ratio of the velocities in the two media."

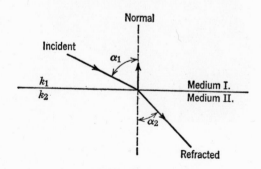

FIGURE #3.1-1

The law is usually given by the equation:

$$k_1 \sin \alpha_1 = k_2 \sin \alpha_2$$

where the index of a medium, k, is determined by $k = c/v$; c is the velocity of light in free space, v the velocity of light in the substance.

Show that Snell's Law has the vector form:

$$k_1 \mathbf{N} \times \mathbf{a}_1 = k_2 \mathbf{N} \times \mathbf{a}_2$$

where $\mathbf{N}$ is a unit normal to the line of division between the two media, $\mathbf{a}_1$ and $\mathbf{a}_2$ are unit vectors along the incident and refracted rays, respectively.

2. Let a rigid body rotate about a fixed axis, say the z-axis. Let $\mathbf{w} = w\mathbf{k}$ be the angular velocity vector. Each point P in the body will move in a circle. Show that the velocity vector associated with P is given by $\mathbf{v} = \mathbf{w} \times \mathbf{r}$, where $\mathbf{r} = OP$, O the origin. (Figure #3.1-2.)

3. The vector moment of the force $\mathbf{F}$ (applied at a point P_1) about the point P_0 is defined as the vector $\mathbf{L}$:

$$\mathbf{L} = \overrightarrow{P_0P_1} \times \mathbf{F}.$$

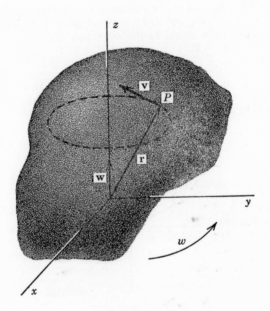

FIGURE #3.1-2

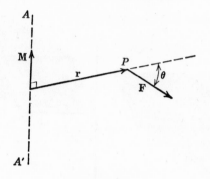

FIGURE #3.1-3

(a) Show that $|\overrightarrow{P_0P_1} \times \mathbf{F}|$ is the scalar moment of the force $\mathbf{F}$ about the axis through P_0 perpendicular to the plane of $\overrightarrow{P_0P_1}$ and $\mathbf{F}$.

(b) Find $\mathbf{L}$ if P_0 is the point $(-1, 2, 1)$ and P_1 is $(3, -1, -1)$ with $\mathbf{F} = -\mathbf{i} + 2\mathbf{j} - 3\mathbf{k}$.

4. The moment of force about an axis can be associated with a vector product by defining a vector moment $\mathbf{M}: \mathbf{M} = \mathbf{r} \times \mathbf{F}$. Show that $|\mathbf{r} \times \mathbf{F}|$ is the scalar moment of force about the axis AA' in Figure #3.1-3. Note that the vectors $\mathbf{r}$ and $\mathbf{F}$ lie in a plane perpendicular to AA'.

3.3 THE TRIPLE PRODUCTS: $\mathbf{a} \cdot \mathbf{b} \times \mathbf{c}$, $\mathbf{a} \times (\mathbf{b} \times \mathbf{c})$, AND $(\mathbf{a} \times \mathbf{b}) \times \mathbf{c}$

Up to this point we have not concerned ourselves with multiple products. Once we have formed the simple products $h\mathbf{a}$, $\mathbf{a} \cdot \mathbf{b}$, $\mathbf{a} \times \mathbf{b}$ it is natural to consider more complicated products, such as $(\mathbf{a} \cdot \mathbf{b})\mathbf{c}$, $\mathbf{a} \cdot \mathbf{b} \times \mathbf{c}$, and $\mathbf{a} \times (\mathbf{b} \times \mathbf{c})$. These latter products, called **triple products,** play an important role in the development of vector analysis and its applications. More elaborate products, such as $(\mathbf{a} \times \mathbf{b}) \cdot (\mathbf{c} \times \mathbf{d})$ and $(\mathbf{a} \times \mathbf{b}) \times (\mathbf{c} \times \mathbf{d})$ can be reduced to simpler forms by means of vector identities. As with trigonometric expressions, the reduction of vector forms by means of vector identities is an important process in the useful applications of vectors to many problems.

The first type of triple product mentioned above, $(\mathbf{a} \cdot \mathbf{b})\mathbf{c}$,

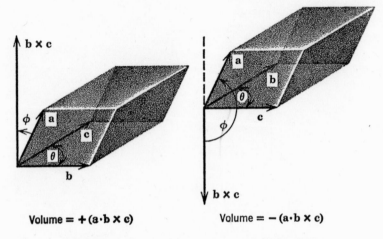

Volume = + (a·b × c)   **Volume = − (a·b × c)**

FIGURE 3.3.1   Geometric interpretation of $\mathbf{a} \cdot \mathbf{b} \times \mathbf{c}$.

does not differ from scalar multiplication for $\mathbf{a} \cdot \mathbf{b} = h$, a scalar, and we have $(\mathbf{a} \cdot \mathbf{b})\mathbf{c} = h\mathbf{c}$.

DEFINITION 3.3.1   *The triple product* $\mathbf{a} \cdot \mathbf{b} \times \mathbf{c}^*$ *is called the* **scalar triple product of a, b,** *and* **c** *in that order.*

The scalar triple product is a scalar quantity, for it is the dot product of the vectors $\mathbf{a}$ and $\mathbf{b} \times \mathbf{c}$. Interpreted geometrically, $\mathbf{a} \cdot \mathbf{b} \times \mathbf{c}$ represents the volume of a parallelepiped with coterminus edges $\mathbf{a}$, $\mathbf{b}$, and $\mathbf{c}$ providing that $\mathbf{a}$, $\mathbf{b}$, and $\mathbf{c}$ are not coplanar.

The geometric interpretation of the scalar triple product is easily verified as follows:

$$
\begin{aligned}
\mathbf{a} \cdot \mathbf{b} \times \mathbf{c} &= |\mathbf{a}| \, |\mathbf{b} \times \mathbf{c}| \cos \phi \\
&= |\mathbf{a}| \, [|\mathbf{b}| \, |\mathbf{c}| \sin \theta] \cos \phi \\
&= [|\mathbf{b}| \, |\mathbf{c}| \sin \theta][|\mathbf{a}| \cos \phi] \\
&= [\text{Area of base}] \begin{bmatrix} + (\text{altitude}) \text{ if } 0 < \phi < \dfrac{\pi}{2} \\[2ex] - (\text{altitude}) \text{ if } \dfrac{\pi}{2} < \phi < \pi \end{bmatrix} \\
&= \pm \text{Volume of the parallelepiped.}
\end{aligned}
$$

* $\mathbf{a} \cdot \mathbf{b} \times \mathbf{c}$ can only mean $\mathbf{a} \cdot (\mathbf{b} \times \mathbf{c})$ since $(\mathbf{a} \cdot \mathbf{b}) \times \mathbf{c}$ is undefined.

Additional properties of the scalar triple product are

THEOREM 3.3.1   [a, b, and c are coplanar] $\Leftrightarrow$ [a · b $\times$ c = 0].

THEOREM 3.3.2   The value of the scalar triple product is not changed by a cyclic permutation of the three elements which form the product:

$$a \cdot b \times c = c \cdot a \times b = b \cdot c \times a.$$

THEOREM 3.3.3   The dot and cross in a scalar triple product can be interchanged without effecting the value of the product:

$$a \cdot b \times c = a \times b \cdot c.$$

THEOREM 3.3.4   If a, b, and c are given in terms of the base vectors i, j, and k, then

$$a \cdot b \times c = \begin{vmatrix} a_x & a_y & a_z \\ b_x & b_y & b_z \\ c_x & c_y & c_z \end{vmatrix}$$

Theorem 3.3.2 is readily established on the basis of the geometric interpretation given to the scalar triple product. Theorem 3.3.3 follows from Theorem 3.3.2 and the commutative property of the scalar product (Algebraic Law 2.1.1). Proofs of Theorems 3.3.1 and 3.3.4 follow.

Proof of THEOREM 3.3.1

Part I.   [a, b, and c are coplanar] $\Rightarrow$ [a · b $\times$ c = 0].

b $\times$ c is perpendicular to both b and c by definition. b $\times$ c is perpendicular to a, for it is given coplanar with b and c. Therefore a · b $\times$ c = 0 by definition of the scalar product.

Part II.   [a · b $\times$ c = 0] $\Rightarrow$ [a, b, and c are coplanar].

a is perpendicular to b $\times$ c, property of the scalar product. But b $\times$ c is also perpendicular to both b and c. Therefore a, b, and c are coplanar, for they are perpendicular to the common vector b $\times$ c.

Proof of THEOREM 3.3.4

$$\mathbf{a} \cdot \mathbf{b} \times \mathbf{c} = (a_x \mathbf{i} + a_y \mathbf{j} + a_z \mathbf{k}) \cdot \begin{vmatrix} \mathbf{i} & \mathbf{j} & \mathbf{k} \\ b_x & b_y & b_z \\ c_x & c_y & c_z \end{vmatrix}$$

$$= (a_x \mathbf{i} + a_y \mathbf{j} + a_z \mathbf{k}) \cdot \left[\begin{vmatrix} b_y & b_z \\ c_y & c_z \end{vmatrix} \mathbf{i} + \begin{vmatrix} b_z & b_x \\ c_z & c_x \end{vmatrix} \mathbf{j} + \begin{vmatrix} b_x & b_y \\ c_x & c_y \end{vmatrix} \mathbf{k} \right]$$

$$= \begin{vmatrix} b_y & b_z \\ c_y & c_z \end{vmatrix} a_x + \begin{vmatrix} b_z & b_x \\ c_z & c_x \end{vmatrix} a_y + \begin{vmatrix} b_x & b_y \\ c_x & c_y \end{vmatrix} a_z$$

$$= \begin{vmatrix} a_x & a_y & a_z \\ b_x & b_y & b_z \\ c_x & c_y & c_z \end{vmatrix} .$$

Note that Theorems 3.3.1, 3.3.2, and 3.3.3 can be proved using Theorem 3.3.4 and the properties of determinants.

The distributive law for the vector product (Algebraic Law 3.1.2) can now be proved rather easily because of the interchangeability of the dot and cross in the scalar triple product.

Proof of the Distributive Law (Algebraic Law 3.1.2)

$$\mathbf{a} \times (\mathbf{b} + \mathbf{c}) = \mathbf{a} \times \mathbf{b} + \mathbf{a} \times \mathbf{c}.$$

Since $[\mathbf{a} \times (\mathbf{b} + \mathbf{c}) = \mathbf{a} \times \mathbf{b} + \mathbf{a} \times \mathbf{c}]$ is equivalent to

$$[\mathbf{a} \times (\mathbf{b} + \mathbf{c}) - \mathbf{a} \times \mathbf{b} - \mathbf{a} \times \mathbf{c} = 0]$$

we let $\qquad \mathbf{u} = \mathbf{a} \times (\mathbf{b} + \mathbf{c}) - \mathbf{a} \times \mathbf{b} - \mathbf{a} \times \mathbf{c}$

and show that $\mathbf{u} = \mathbf{0}$.

Let $\mathbf{v}$ be an arbitrary vector and form the scalar product

$$\mathbf{v} \cdot \mathbf{u} = \mathbf{v} \cdot [\mathbf{a} \times (\mathbf{b} + \mathbf{c}) - \mathbf{a} \times \mathbf{b} - \mathbf{a} \times \mathbf{c}]$$
$$= \mathbf{v} \cdot [\mathbf{a} \times (\mathbf{b} + \mathbf{c})] - \mathbf{v} \cdot \mathbf{a} \times \mathbf{b} - \mathbf{v} \cdot \mathbf{a} \times \mathbf{c}$$

(since the scalar product is distributive).

$$= \mathbf{v} \times \mathbf{a} \cdot (\mathbf{b} + \mathbf{c}) - \mathbf{v} \times \mathbf{a} \cdot \mathbf{b} - \mathbf{v} \times \mathbf{a} \cdot \mathbf{c} \quad \text{(Theorem 3.3.3)}$$
$$= \mathbf{v} \times \mathbf{a} \cdot \mathbf{b} + \mathbf{v} \times \mathbf{a} \cdot \mathbf{c} - \mathbf{v} \times \mathbf{a} \cdot \mathbf{b} - \mathbf{v} \times \mathbf{a} \cdot \mathbf{c}$$
$$= 0$$

Hence, either $\mathbf{v} = \mathbf{0}$, $\mathbf{u} = \mathbf{0}$, or $\mathbf{v}$ is perpendicular to $\mathbf{u}$. But since $\mathbf{v}$ is an arbitrary vector, $\mathbf{u} = \mathbf{0}$.

DEFINITION 3.3.2   *The triple products* $\mathbf{a} \times (\mathbf{b} \times \mathbf{c})$ *and* $(\mathbf{a} \times \mathbf{b}) \times \mathbf{c}$ *are called* **vector triple products.**

The vector triple products are vector products of vectors and hence are vector quantities. The vector $\mathbf{a} \times (\mathbf{b} \times \mathbf{c})$ is perpendicular to $\mathbf{a}$ and to $\mathbf{b} \times \mathbf{c}$. The vector $\mathbf{b} \times \mathbf{c}$ is perpendicular to $\mathbf{b}$ and to $\mathbf{c}$. Thus the vectors $\mathbf{a} \times (\mathbf{b} \times \mathbf{c})$, $\mathbf{b}$, and $\mathbf{c}$ must be coplanar.

THEOREM 3.3.5   There exist scalars m and n such that

$$\mathbf{a} \times (\mathbf{b} \times \mathbf{c}) = m\mathbf{b} + n\mathbf{c}.$$

THEOREM 3.3.6   There exist scalars p and q such that

$$(\mathbf{a} \times \mathbf{b}) \times \mathbf{c} = p\mathbf{a} + q\mathbf{b}.$$

From Theorems 3.3.5 and 3.3.6 it is clear that

$$\mathbf{a} \times (\mathbf{b} \times \mathbf{c}) \neq (\mathbf{a} \times \mathbf{b}) \times \mathbf{c}$$

so that the parentheses in the vector triple product are necessary and cannot be removed or shifted without changing the vector product.

Vector triple products are generally evaluated by means of the following two vector identities.

IDENTITY 3.3.1 $\mathbf{a} \times (\mathbf{b} \times \mathbf{c}) = (\mathbf{a} \cdot \mathbf{c})\mathbf{b} - (\mathbf{a} \cdot \mathbf{b})\mathbf{c}.$

IDENTITY 3.3.2 $(\mathbf{a} \times \mathbf{b}) \times \mathbf{c} = (\mathbf{c} \cdot \mathbf{a})\mathbf{b} - (\mathbf{c} \cdot \mathbf{b})\mathbf{a}.$

A detailed proof of Identity 3.3.1 is given below, for it is instructive to observe the manner in which the scalar and vector properties are used. In proving the identity we start with

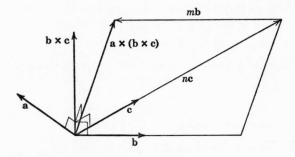

FIGURE 3.3.2 $\mathbf{a} \times (\mathbf{b} \times \mathbf{c}) = m\mathbf{b} + n\mathbf{c}.$

Theorem 3.3.5 and show that $m = -(\mathbf{a} \cdot \mathbf{b})$ and $n = \mathbf{a} \cdot \mathbf{c}$. Identity 3.3.2 follows directly from Identity 3.3.1 with the application of the anticommutative law, $\mathbf{a} \times (\mathbf{b} \times \mathbf{c}) = -[(\mathbf{b} \times \mathbf{c}) \times \mathbf{a}]$.

Proof of IDENTITY 3.3.1

$$\mathbf{a} \times (\mathbf{b} \times \mathbf{c}) = m\mathbf{b} + n\mathbf{c} \qquad \text{(Theorem 3.3.5)}$$
$$\mathbf{a} \cdot [\mathbf{a} \times (\mathbf{b} \times \mathbf{c})] = m(\mathbf{a} \cdot \mathbf{b}) + n(\mathbf{a} \cdot \mathbf{c})$$
$$(\mathbf{a} \times \mathbf{a}) \cdot (\mathbf{b} \times \mathbf{c}) = m(\mathbf{a} \cdot \mathbf{b}) + n(\mathbf{a} \cdot \mathbf{c}) \qquad \text{(Theorem 3.3.3)}$$
$$\mathbf{0} \cdot (\mathbf{b} \times \mathbf{c}) = m(\mathbf{a} \cdot \mathbf{b}) + n(\mathbf{a} \cdot \mathbf{c})$$
$$0 = m(\mathbf{a} \cdot \mathbf{b}) + n(\mathbf{a} \cdot \mathbf{c})$$

Therefore $\dfrac{m}{\mathbf{a} \cdot \mathbf{c}} = -\dfrac{n}{\mathbf{a} \cdot \mathbf{b}} = R$ a scalar,

$$m = R(\mathbf{a} \cdot \mathbf{c}) \text{ and } n = -R(\mathbf{a} \cdot \mathbf{b}).$$

Hence $\qquad \mathbf{a} \times (\mathbf{b} \times \mathbf{c}) = R[(\mathbf{a} \cdot \mathbf{c})\mathbf{b} - (\mathbf{a} \cdot \mathbf{b})\mathbf{c}]. \qquad (i)$

Now

$$\mathbf{b} \cdot [\mathbf{a} \times (\mathbf{b} \times \mathbf{c})] = R[(\mathbf{a} \cdot \mathbf{c})(\mathbf{b} \cdot \mathbf{b}) - (\mathbf{a} \cdot \mathbf{b})(\mathbf{b} \cdot \mathbf{c})]$$
$$\mathbf{b} \cdot [\mathbf{a} \times (\mathbf{b} \times \mathbf{c})] = (\mathbf{b} \times \mathbf{a}) \cdot (\mathbf{b} \times \mathbf{c}) \qquad \text{(Theorem 3.3.3)}$$
$$= -(\mathbf{a} \times \mathbf{b}) \cdot (\mathbf{b} \times \mathbf{c}) \qquad \text{(Algebraic Law 3.1.1)}$$
$$= -\mathbf{a} \cdot [\mathbf{b} \times (\mathbf{b} \times \mathbf{c})] \qquad \text{(Theorem 3.3.3)}$$

Hence $-\mathbf{a} \cdot [\mathbf{b} \times (\mathbf{b} \times \mathbf{c})] = R[(\mathbf{a} \cdot \mathbf{c})(\mathbf{b} \cdot \mathbf{b}) - (\mathbf{a} \cdot \mathbf{b})(\mathbf{b} \cdot \mathbf{c})]. \quad (ii)$

Now we prove the following lemma: $\mathbf{b} \times (\mathbf{b} \times \mathbf{c}) = (\mathbf{b} \cdot \mathbf{c})\mathbf{b} - b^2\mathbf{c}$.

$$\mathbf{b} \times (\mathbf{b} \times \mathbf{c}) = R[(\mathbf{b} \cdot \mathbf{c})\mathbf{b} - b^2\mathbf{c}] \qquad \text{(see (i))}$$
$$\mathbf{c} \cdot [\mathbf{b} \times (\mathbf{b} \times \mathbf{c})] = R[(\mathbf{b} \cdot \mathbf{c})^2 - b^2c^2]$$
$$(\mathbf{c} \times \mathbf{b}) \cdot (\mathbf{b} \times \mathbf{c}) = R[b^2c^2 \cos^2 \theta - b^2c^2]$$
$$-(\mathbf{b} \times \mathbf{c})^2 = -Rb^2c^2 \sin^2 \theta.$$

Therefore $R = 1$ and $\mathbf{b} \times (\mathbf{b} \times \mathbf{c}) = (\mathbf{b} \cdot \mathbf{c})\mathbf{b} - b^2\mathbf{c}$.

Then by (ii) and the lemma, we have

$$-\mathbf{a} \cdot [(\mathbf{b} \cdot \mathbf{c})\mathbf{b} - b^2\mathbf{c}] = R[(\mathbf{a} \cdot \mathbf{c})(\mathbf{b} \cdot \mathbf{b}) - (\mathbf{a} \cdot \mathbf{b})(\mathbf{b} \cdot \mathbf{c})]$$
$$-(\mathbf{a} \cdot \mathbf{b})(\mathbf{b} \cdot \mathbf{c}) + b^2(\mathbf{a} \cdot \mathbf{c}) = R[(\mathbf{a} \cdot \mathbf{c})b^2 - (\mathbf{a} \cdot \mathbf{b})(\mathbf{b} \cdot \mathbf{c})].$$

Hence in (ii) $R = 1$ and we have proved that

$$\mathbf{a} \times (\mathbf{b} \times \mathbf{c}) = (\mathbf{a} \cdot \mathbf{c})\mathbf{b} - (\mathbf{a} \cdot \mathbf{b})\mathbf{c}.$$

Note that the above proof is independent of any coordinate system. Identity 3.3.1 can also be established by expressing **a**, **b**, and **c** in terms of the base vectors **i**, **j**, and **k** and showing that the corresponding components of the left- and right-hand members of $\mathbf{a} \times (\mathbf{b} \times \mathbf{c}) = (\mathbf{a} \cdot \mathbf{c})\mathbf{b} - (\mathbf{a} \cdot \mathbf{b})\mathbf{c}$ are equal.

3.4 VECTOR IDENTITIES

In addition to Identities 3.3.1 and 3.3.2, two further identities of frequent use are

IDENTITY 3.4.1 $\quad (\mathbf{a} \times \mathbf{b}) \cdot (\mathbf{c} \times \mathbf{d}) = \begin{vmatrix} \mathbf{a} \cdot \mathbf{c} & \mathbf{a} \cdot \mathbf{d} \\ \mathbf{b} \cdot \mathbf{c} & \mathbf{b} \cdot \mathbf{d} \end{vmatrix}.$

IDENTITY 3.4.2 $\quad (\mathbf{a} \times \mathbf{b}) \times (\mathbf{c} \times \mathbf{d}) = (\mathbf{a} \cdot \mathbf{c} \times \mathbf{d})\mathbf{b} - (\mathbf{b} \cdot \mathbf{c} \times \mathbf{d})\mathbf{a}$
$$= (\mathbf{a} \cdot \mathbf{b} \times \mathbf{d})\mathbf{c} - (\mathbf{a} \cdot \mathbf{b} \times \mathbf{c})\mathbf{d}.$$

Proof of IDENTITY 3.4.1

$$\begin{aligned}
(\mathbf{a} \times \mathbf{b}) \cdot (\mathbf{c} \times \mathbf{d}) &= \mathbf{a} \times \mathbf{b} \cdot (\mathbf{c} \times \mathbf{d}) \\
&= \mathbf{a} \cdot \mathbf{b} \times (\mathbf{c} \times \mathbf{d}) \\
&= \mathbf{a} \cdot [(\mathbf{b} \cdot \mathbf{d})\mathbf{c} - (\mathbf{b} \cdot \mathbf{c})\mathbf{d}] \\
&= (\mathbf{b} \cdot \mathbf{d})(\mathbf{a} \cdot \mathbf{c}) - (\mathbf{b} \cdot \mathbf{c})(\mathbf{a} \cdot \mathbf{d}) \\
&= \begin{vmatrix} \mathbf{a} \cdot \mathbf{c} & \mathbf{a} \cdot \mathbf{d} \\ \mathbf{b} \cdot \mathbf{c} & \mathbf{b} \cdot \mathbf{d} \end{vmatrix}.
\end{aligned}$$

Proof of IDENTITY 3.4.2

Let $\mathbf{p} = \mathbf{c} \times \mathbf{d}$; then

$$(\mathbf{a} \times \mathbf{b}) \times (\mathbf{c} \times \mathbf{d}) = (\mathbf{a} \times \mathbf{b}) \times \mathbf{p}.$$

But $\quad (\mathbf{a} \times \mathbf{b}) \times \mathbf{p} = (\mathbf{a} \cdot \mathbf{p})\mathbf{b} - (\mathbf{b} \cdot \mathbf{p})\mathbf{a}$
$$= (\mathbf{a} \cdot \mathbf{c} \times \mathbf{d})\mathbf{b} - (\mathbf{b} \cdot \mathbf{c} \times \mathbf{d})\mathbf{a}.$$

Let $\mathbf{q} = \mathbf{a} \times \mathbf{b}$; then

$$(\mathbf{a} \times \mathbf{b}) \times (\mathbf{c} \times \mathbf{d}) = \mathbf{q} \times (\mathbf{c} \times \mathbf{d}).$$

But $\quad \mathbf{q} \times (\mathbf{c} \times \mathbf{d}) = (\mathbf{q} \cdot \mathbf{d})\mathbf{c} - (\mathbf{q} \cdot \mathbf{c})\mathbf{d}$
$$= (\mathbf{a} \times \mathbf{b} \cdot \mathbf{d})\mathbf{c} - (\mathbf{a} \times \mathbf{b} \cdot \mathbf{c})\mathbf{d}.$$

Thus we have the two results:

$$(\mathbf{a} \times \mathbf{b}) \times (\mathbf{c} \times \mathbf{d}) = (\mathbf{a} \cdot \mathbf{c} \times \mathbf{d})\mathbf{b} - (\mathbf{b} \cdot \mathbf{c} \times \mathbf{d})\mathbf{a}$$
$$= (\mathbf{a} \cdot \mathbf{b} \times \mathbf{d})\mathbf{c} - (\mathbf{a} \cdot \mathbf{b} \times \mathbf{c})\mathbf{d}.$$

Note that in the above results the conclusion that

$$(\mathbf{a} \cdot \mathbf{c} \times \mathbf{d})\mathbf{b} - (\mathbf{b} \cdot \mathbf{c} \times \mathbf{d})\mathbf{a} = (\mathbf{a} \cdot \mathbf{b} \times \mathbf{d})\mathbf{c} - (\mathbf{a} \cdot \mathbf{b} \times \mathbf{c})\mathbf{d}$$

asserts that for any four vectors **a**, **b**, **c**, and **d** in three space we can obtain a unique linear representation of any one of the vectors in terms of the other three if the three are not coplanar.

No attempt should be made to memorize the Identities 3.4.1 and 3.4.2. One should, however, try to develop a facility in applying them. The examples that follow will illustrate how these identities and others considered earlier can be used to simplify vector expressions or establish new identities.

EXAMPLE 3.4.1   Simplify $(a + b) \cdot (b + c) \times (c + a)$.

Solution

$$
\begin{aligned}
(a + b) &\cdot (b + c) \times (c + a) \\
&= (a + b) \cdot [(b + c) \times c + (b + c) \times a] \\
&= (a + b) \cdot [b \times c + c \times c + b \times a + c \times a] \\
&= a \cdot b \times c + a \cdot b \times a + a \cdot c \times a + b \cdot b \times c \\
&\qquad\qquad\qquad\qquad + b \cdot b \times a + b \cdot c \times a \\
&= 2(a \cdot b \times c).
\end{aligned}
$$

EXAMPLE 3.4.2   Show that

$$a \times (b \times c) + b \times (c \times a) + c \times (a \times b) = 0.$$

Solution

$$
\begin{aligned}
a \times (b \times c) &+ b \times (c \times a) + c \times (a \times b) \\
&= [(a \cdot c)b - (a \cdot b)c] + [(b \cdot a)c - (b \cdot c)a] + [(c \cdot b)a - (c \cdot a)b] \\
&= 0.
\end{aligned}
$$

EXAMPLE 3.4.3   Show that $(a \times b) \cdot (b \times c) \times (c \times a) = (a \cdot b \times c)^2$.

Solution

$$
\begin{aligned}
(a \times b) \cdot (b \times c) \times (c \times a) &= a \times b \cdot [(b \cdot c \times a)c - (c \cdot c \times a)b] \\
&= a \times b \cdot [(b \cdot c \times a)c] \\
&= (b \cdot c \times a)(a \times b \cdot c) \\
&= (b \cdot c \times a)(c \cdot a \times b) \\
&= (b \cdot c \times a)(c \times a \cdot b) \\
&= (b \cdot c \times a)^2 \\
&= (a \cdot b \times c)^2.
\end{aligned}
$$

SUMMARY

3.3   The Triple Products

DEFINITION 3.3.1   *The triple product* $a \cdot b \times c$ *is called the* **scalar triple product** *of* **a**, **b**, *and* **c** *in that order.*

THEOREM 3.3.1 [a, b, and c are coplanar] $\Leftrightarrow$ [$\mathbf{a} \cdot \mathbf{b} \times \mathbf{c} = 0$].

THEOREM 3.3.2 The value of the scalar triple product is not changed by a cyclic permutation of the three elements which form the product:

$$\mathbf{a} \cdot \mathbf{b} \times \mathbf{c} = \mathbf{c} \cdot \mathbf{a} \times \mathbf{b} = \mathbf{b} \cdot \mathbf{c} \times \mathbf{a}.$$

THEOREM 3.3.3 The dot and cross in a scalar triple product can be interchanged without effecting the value of the product:

$$\mathbf{a} \cdot \mathbf{b} \times \mathbf{c} = \mathbf{a} \times \mathbf{b} \cdot \mathbf{c}.$$

THEOREM 3.3.4 If a, b, and c are given in terms of the base vectors i, j, and k, then

$$\mathbf{a} \cdot \mathbf{b} \times \mathbf{c} = \begin{vmatrix} a_x & a_y & a_z \\ b_x & b_y & b_z \\ c_x & c_y & c_z \end{vmatrix}.$$

DEFINITION 3.3.2 *The triple products* $\mathbf{a} \times (\mathbf{b} \times \mathbf{c})$ *and* $(\mathbf{a} \times \mathbf{b}) \times \mathbf{c}$ *are called* **vector** triple products.

THEOREM 3.3.5 There exist scalars m and n such that

$$\mathbf{a} \times (\mathbf{b} \times \mathbf{c}) = m\mathbf{b} + n\mathbf{c}.$$

THEOREM 3.3.6 There exist scalars p and q such that

$$(\mathbf{a} \times \mathbf{b}) \times \mathbf{c} = p\mathbf{a} + q\mathbf{b}.$$

IDENTITY 3.3.1 $\mathbf{a} \times (\mathbf{b} \times \mathbf{c}) = (\mathbf{a} \cdot \mathbf{c})\mathbf{b} - (\mathbf{a} \cdot \mathbf{b})\mathbf{c}$.

IDENTITY 3.3.2 $(\mathbf{a} \times \mathbf{b}) \times \mathbf{c} = (\mathbf{c} \cdot \mathbf{a})\mathbf{b} - (\mathbf{c} \cdot \mathbf{b})\mathbf{a}$.

3.4 Vector Identities

IDENTITY 3.4.1 $(\mathbf{a} \times \mathbf{b}) \cdot (\mathbf{c} \times \mathbf{d}) = \begin{vmatrix} \mathbf{a} \cdot \mathbf{c} & \mathbf{a} \cdot \mathbf{d} \\ \mathbf{b} \cdot \mathbf{c} & \mathbf{b} \cdot \mathbf{d} \end{vmatrix}$.

IDENTITY 3.4.2 $(\mathbf{a} \times \mathbf{b}) \times (\mathbf{c} \times \mathbf{d}) = (\mathbf{a} \cdot \mathbf{c} \times \mathbf{d})\mathbf{b} - (\mathbf{b} \cdot \mathbf{c} \times \mathbf{d})\mathbf{a}$
$$= (\mathbf{a} \cdot \mathbf{b} \times \mathbf{d})\mathbf{c} - (\mathbf{a} \cdot \mathbf{b} \times \mathbf{c})\mathbf{d}.$$

EXAMPLE 3.4.1 Simplify $(\mathbf{a} + \mathbf{b}) \cdot (\mathbf{b} + \mathbf{c}) \times (\mathbf{c} + \mathbf{a})$.

EXAMPLE 3.4.2 Show that $\mathbf{a} \times (\mathbf{b} \times \mathbf{c}) + \mathbf{b} \times (\mathbf{c} \times \mathbf{a}) + \mathbf{c} \times (\mathbf{a} \times \mathbf{b}) = \mathbf{0}$.

EXAMPLE 3.4.3 Show that $(\mathbf{a} \times \mathbf{b}) \cdot (\mathbf{b} \times \mathbf{c}) \times (\mathbf{c} \times \mathbf{a}) = (\mathbf{a} \cdot \mathbf{b} \times \mathbf{c})^2$

PROBLEM SET #3.2

A. General

1. Prove Theorem 3.3.1, using Theorem 3.3.4 and the properties of determinants.

2. Prove Theorem 3.3.2, using Theorem 3.3.4 and the properties of determinants.

3. Prove Theorem 3.3.3, using Theorem 3.3.4 and the properties of determinants.

4. Prove Identity 3.3.1, assuming $\mathbf{a}$, $\mathbf{b}$, and $\mathbf{c}$ are given in terms of the base vectors $\mathbf{i}$, $\mathbf{j}$, and $\mathbf{k}$, by showing that the corresponding components of the left- and right-hand sides of $\mathbf{a} \times (\mathbf{b} \times \mathbf{c}) = (\mathbf{a} \cdot \mathbf{c})\mathbf{b} - (\mathbf{a} \cdot \mathbf{b})\mathbf{c}$ are equal.

5. Given $\mathbf{a} = 2\mathbf{i} + \mathbf{j} - 3\mathbf{k}$, $\mathbf{b} = \mathbf{i} - 3\mathbf{j} - \mathbf{k}$, $\mathbf{c} = -2\mathbf{i} + 2\mathbf{j} + \mathbf{k}$, find

    (a) $\mathbf{a} \cdot \mathbf{b} \times \mathbf{c}$ (b) $\mathbf{b} \times \mathbf{a} \cdot \mathbf{c}$ (c) $\mathbf{c} \cdot \mathbf{a} \times \mathbf{b}$
    (d) $\mathbf{a} \times (\mathbf{b} \times \mathbf{c})$ (e) $(\mathbf{a} \times \mathbf{b}) \times \mathbf{c}$ (f) $\mathbf{b} \times (\mathbf{a} \times \mathbf{c})$
    (g) $(\mathbf{a} \times \mathbf{b}) \cdot (\mathbf{b} \times \mathbf{c})$ (h) $(\mathbf{c} \cdot \mathbf{b})(\mathbf{a} \times \mathbf{c})$ (i) $(\mathbf{a} \times \mathbf{b}) \times (\mathbf{b} \times \mathbf{c})$

6. Given $\mathbf{a} = \mathbf{i} - \mathbf{j} + 2\mathbf{k}$, $\mathbf{b} = 2\mathbf{i} - 3\mathbf{k}$, $\mathbf{c} = 2\mathbf{j} - \mathbf{k}$, find the quantities (a) through (i) of problem 5 above.

7. Show that

$$\mathbf{i} = \frac{\mathbf{j} \times \mathbf{k}}{(\mathbf{i} \cdot \mathbf{j} \times \mathbf{k})}, \quad \mathbf{j} = \frac{\mathbf{k} \times \mathbf{i}}{(\mathbf{i} \cdot \mathbf{j} \times \mathbf{k})}, \quad \mathbf{k} = \frac{\mathbf{i} \times \mathbf{j}}{(\mathbf{i} \cdot \mathbf{j} \times \mathbf{k})}$$

8. Show that $[\mathbf{a} \times (\mathbf{b} \times \mathbf{c}) = (\mathbf{a} \times \mathbf{b}) \times \mathbf{c}] \Leftrightarrow [\mathbf{b} \times (\mathbf{c} \times \mathbf{a}) = \mathbf{0}]$.

9. Given that $\mathbf{a} \times \mathbf{b} \neq \mathbf{0}$ and $\mathbf{a}$, $\mathbf{b}$, and $\mathbf{c}$ are coplanar, show that

$$\mathbf{c} = \frac{\begin{vmatrix} \mathbf{c} \cdot \mathbf{a} & \mathbf{a} \cdot \mathbf{b} \\ \mathbf{c} \cdot \mathbf{b} & \mathbf{b} \cdot \mathbf{b} \end{vmatrix}}{\begin{vmatrix} \mathbf{a} \cdot \mathbf{a} & \mathbf{a} \cdot \mathbf{b} \\ \mathbf{a} \cdot \mathbf{b} & \mathbf{b} \cdot \mathbf{b} \end{vmatrix}} \mathbf{a} + \frac{\begin{vmatrix} \mathbf{a} \cdot \mathbf{a} & \mathbf{c} \cdot \mathbf{a} \\ \mathbf{a} \cdot \mathbf{b} & \mathbf{c} \cdot \mathbf{b} \end{vmatrix}}{\begin{vmatrix} \mathbf{a} \cdot \mathbf{a} & \mathbf{a} \cdot \mathbf{b} \\ \mathbf{a} \cdot \mathbf{b} & \mathbf{b} \cdot \mathbf{b} \end{vmatrix}} \mathbf{b}.$$

10. It can be shown that three vectors $\mathbf{a}$, $\mathbf{b}$, and $\mathbf{c}$ form a basis for the vectors in 3-space if and only if $\mathbf{a} \cdot \mathbf{b} \times \mathbf{c} \neq 0$. (See Theorem 3.3.1.) If $\mathbf{a} \cdot \mathbf{b} \times \mathbf{c} \neq 0$, then show that an arbitrary vector $\mathbf{v}$ in 3-space can be represented as a linear combination of $\mathbf{a}$, $\mathbf{b}$, and $\mathbf{c}$ as follows:

$$\mathbf{v} = \left(\frac{\mathbf{v} \cdot \mathbf{b} \times \mathbf{c}}{\mathbf{a} \cdot \mathbf{b} \times \mathbf{c}}\right) \mathbf{a} + \left(\frac{\mathbf{v} \cdot \mathbf{c} \times \mathbf{a}}{\mathbf{a} \cdot \mathbf{b} \times \mathbf{c}}\right) \mathbf{b} + \left(\frac{\mathbf{v} \cdot \mathbf{a} \times \mathbf{b}}{\mathbf{a} \cdot \mathbf{b} \times \mathbf{c}}\right) \mathbf{c}.$$

11. Prove that $\mathbf{a} \times [\mathbf{b} \times (\mathbf{c} \times \mathbf{d})] = (\mathbf{b} \cdot \mathbf{d})(\mathbf{a} \times \mathbf{c}) - (\mathbf{b} \cdot \mathbf{c})(\mathbf{a} \times \mathbf{d})$.

12. Show that

$$(\mathbf{a} \cdot \mathbf{b} \times \mathbf{c})(\mathbf{p} \cdot \mathbf{q} \times \mathbf{r}) = \begin{vmatrix} \mathbf{a} \cdot \mathbf{p} & \mathbf{a} \cdot \mathbf{q} & \mathbf{a} \cdot \mathbf{r} \\ \mathbf{b} \cdot \mathbf{p} & \mathbf{b} \cdot \mathbf{q} & \mathbf{b} \cdot \mathbf{r} \\ \mathbf{c} \cdot \mathbf{p} & \mathbf{c} \cdot \mathbf{q} & \mathbf{c} \cdot \mathbf{r} \end{vmatrix}.$$

13. (a) Prove that

$$(\mathbf{b} \times \mathbf{c}) \cdot (\mathbf{a} \times \mathbf{d}) + (\mathbf{c} \times \mathbf{a}) \cdot (\mathbf{b} \times \mathbf{d}) + (\mathbf{a} \times \mathbf{b}) \cdot (\mathbf{c} \times \mathbf{d}) = 0.$$

(b) Use the results of part (a) to show that

$$\sin (\alpha + \beta) \sin (\alpha - \beta) = \sin^2 \alpha - \sin^2 \beta.$$

14. Show that if $\mathbf{a}$, $\mathbf{b}$, $\mathbf{c}$, and $\mathbf{d}$ are coplanar, then

$$(\mathbf{a} \times \mathbf{b}) \times (\mathbf{c} \times \mathbf{d}) = \mathbf{0}.$$

Is the converse true? If so, prove it; if not, find a counterexample.

15. Write the system:

$$x - y + 2z = 3$$
$$2x + y - z = 7$$
$$3y - 2z = 4$$

as the single vector equation $x\mathbf{a} + y\mathbf{b} + z\mathbf{c} = \mathbf{d}$, where $\mathbf{a} = \mathbf{i} + 2\mathbf{j}$, $\mathbf{b} = -\mathbf{i} + \mathbf{j} + 3\mathbf{k}$, $\mathbf{c} = 2\mathbf{i} - \mathbf{j} - 2\mathbf{k}$, $\mathbf{d} = 3\mathbf{i} + 7\mathbf{j} + 4\mathbf{k}$. To solve for x, dot both sides of the vector equation with $\mathbf{b} \times \mathbf{c}$ to obtain

$$(\mathbf{a} \cdot \mathbf{b} \times \mathbf{c})x + (\mathbf{b} \cdot \mathbf{b} \times \mathbf{c})y + (\mathbf{c} \cdot \mathbf{b} \times \mathbf{c})z = \mathbf{d} \cdot \mathbf{b} \times \mathbf{c}.$$

Also solve for y and z, using a similar procedure.

16. Given $\mathbf{a} = \mathbf{i} - \mathbf{j} + \mathbf{k}$ and $\mathbf{b} = \mathbf{j} + \mathbf{k}$, express $\mathbf{b}$ as a linear combination of vectors parallel to and perpendicular to $\mathbf{a}$ in the plane determined by $\mathbf{a}$ and $\mathbf{b}$ when taken from a common point. (*Hint:* What is the direction of $\mathbf{a} \times (\mathbf{a} \times \mathbf{b})$ with respect to $\mathbf{a}$?)

B. Geometric

1. Find the volume of a parallelepiped having coterminus edges $\mathbf{a} = \mathbf{i} - 2\mathbf{j} + 3\mathbf{k}$, $\mathbf{b} = \mathbf{j} + 3\mathbf{k}$, and $\mathbf{c} = \mathbf{i} - \mathbf{k}$.

2. Find the volume of a tetrahedron with coterminus edges $\mathbf{a} = \mathbf{i} - 2\mathbf{j} + 3\mathbf{k}$, $\mathbf{b} = \mathbf{j} + 3\mathbf{k}$, and $\mathbf{c} = \mathbf{i} - \mathbf{k}$.

3. Write the equation of a plane through the three points P_1, P_2, and P_3 in the form of a scalar triple product set equal to zero.

4. Use the results of Problem 3 to write the equation of a plane (in scalar form) through the points $P_1(-1, -2, 3)$, $P_2(0, 1, -2)$, and $P_3(2, 0, 0)$.

5. Replace the three points in Problem 4 with $P_1(0, 1, 5)$, $P_2(1, 1, 1)$, $P_3(1, -1, 0)$ and solve.

6. Replace the three points in Problem 4 with $P_1(2, -5, 7)$, $P_2(0, 0, -8)$, $P_3(1, 1, 0)$ and solve.

7. Let a plane pass through the three points P_1, P_2, and P_3. Find a vector equation for the distance between the plane and a point P_4 not on the plane.

8. Use the results of Problem 7 to find the distance from $P_4(1, -1, 3)$ to the plane passing through the three points $P_1(2, 1, 1)$, $P_2(0, -3, 1)$, and $P_3(-1, 1, 5)$.

9. Replace the four points in Problem 8 with $P_4(-3, -2, 7)$, $P_1(1, 1, -1)$, $P_2(2, 3, -1)$, and $P_3(-2, 5, 1)$.

10. Find the distance between a point $P(1, 5, -2)$ and the plane $2x - 3y + 3z - 6 = 0$.

11. Let P_1, P_2, P_3, and P_4 be four noncoplanar points such that the lines through P_1P_2 and P_3P_4 do not intersect. Write a vector equation for the distance between the two nonintersecting skew lines.

12. Using the results of Problem 11, find the distance (a numerical value) between $\overline{P_1P_2}$ and $\overline{P_3P_4}$ given that $P_1(0, 0, 1)$, $P_2(-1, 2, -1)$, $P_3(1, 2, 1)$, and $P_4(0, -1, -1)$.

13. Replace the four points in Problem 12 with $P_1(1, 2, -1)$, $P_2(3, -1, 5)$, $P_3(2, 4, -6)$, and $P_4(0, 0, 0)$. Find the numerical distance between $\overline{P_1P_2}$ and $\overline{P_3P_4}$.

14. Find the distance between the two nonintersecting lines:

$$m_1: x = 2t, \ y = 2 - t, \ z = 5 + 3t$$
$$m_2: x = 1 + t, \ y = 1 - t, \ z = 2t$$

C. Physics

1. Show that the vector moment M about an axis AA' of a force $\mathbf{F}$ (not, in general, coplanar with AA') applied at a point P_1 is $\mathbf{M} = (\mathbf{N} \times \overrightarrow{P_0P_1} \cdot \mathbf{F})\mathbf{N}$, where $\mathbf{N}$ is a unit vector along AA' and P_0 is the projection of P_1 on AA' (see Figure #3.2-1).

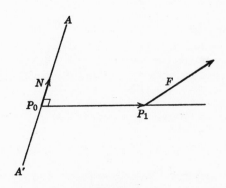

FIGURE #3.2-1

2. Using the result of Problem 1, find the vector moment **M** if the axis AA' is the y-axis and the force $\mathbf{F} = 20\mathbf{i} + 32\mathbf{j} - 4\mathbf{k}$ is applied at $P_1(2, -4, 3)$.

3. Using the result of Problem 1, find the vector moment **M** if the axis AA' is the x-axis and the force $\mathbf{F} = 100\mathbf{i} - 20\mathbf{k}$ is applied at $P_1(3, 0, -1)$.

4. Let an object be located in a Cartesian coordinate system with the forces $\mathbf{F}_1 = \mathbf{i} + 2\mathbf{j} - 3\mathbf{k}$ and $\mathbf{F}_2 = 2\mathbf{i} - 5\mathbf{j}$ applied at $P_1(3, 1, 2)$ and $P_2(-1, 1, 2)$, respectively. Find a force $\mathbf{F}_3$, if it exists, so that when it is applied at the point $P_3(-2, -1, 1)$, no rotation will occur around the z-axis, that is, so that $\mathbf{M}_1 + \mathbf{M}_2 + \mathbf{M}_3 = 0$.

5. Let an object be located in a Cartesian coordinate system with the forces $\mathbf{F}_1 = 2\mathbf{j} + 5\mathbf{k}$ and $\mathbf{F}_2 = 6\mathbf{i} - 2\mathbf{k}$ applied at $P_1(1, -1, 5)$ and $P_2(1, 1, 1)$, respectively. Find a force $\mathbf{F}_3$, if it exists, so that when it is applied at the point $P_3(-6, 7, 3)$, no rotation will occur around the x-axis.

6. For a body to be in equilibrium the vector equations $\mathbf{F} = \mathbf{0}$ and $\mathbf{M} = \mathbf{0}$ must be satisfied, where **F** is the resultant of all external forces acting on the body and **M** is the resultant vector moment of all external forces about any axis. Find the force $\mathbf{F}_3$, if it exists, for the object in Problem 4 to be in equilibrium. (*Hint:* $[\mathbf{F} = \mathbf{0}] \Leftrightarrow [F_x = 0, F_y = 0, F_z = 0]$ and $[\mathbf{M} = \mathbf{0}] \Leftrightarrow [M_x = 0, M_y = 0, M_z = 0]$.)

7. Find the force $\mathbf{F}_3$, if it exists, for the object in Problem 5 to be in equilibrium. (See Problem 6.)

4

vector equations
curves and surfaces

4.1 VECTOR EQUATIONS

A variety of vector equations have been encountered through-
out the preceding sections of our discussion of vectors. For
example,

(a) If a point P divides a segment AB so that $PB/PA = \alpha/\beta$, then for any point O we have

$$\overrightarrow{OP} = \frac{\alpha}{\alpha + \beta}\,\mathbf{a} + \frac{\beta}{\alpha + \beta}\,\mathbf{b}$$

where $\mathbf{a} = \overrightarrow{OA}$ and $\mathbf{b} = \overrightarrow{OB}$.   (See Problem Set #1, B-3.)

(b) In describing loci we have considered vector equations for
the locus of all points P such that for A and B fixed points

(i) $|\overrightarrow{AP}| = |\overrightarrow{BP}|$ (ii) $|\overrightarrow{AB}| + |\overrightarrow{BP}| = |\overrightarrow{AP}|$

(iii) $(\overrightarrow{AB} - \overrightarrow{AP}) \cdot \overrightarrow{AB} = 0$ (iv) $\overrightarrow{AB} \times \overrightarrow{AP} = \mathbf{0}$

(See Problem Set #1, B-11; #2.1, B-7; #3.1, B-3.)

(c) We have used vector equations to derive the equations of lines, $\mathbf{n} \times \overrightarrow{PQ} = \mathbf{0}$; planes, $\mathbf{n} \cdot \overrightarrow{PQ} = 0$; distance from a point to a line, $|\mathbf{n}|d = \mathbf{n} \cdot \overrightarrow{QP}$; volume of a parallelepiped, $\mathbf{a} \times \mathbf{b} \cdot \mathbf{c} = V$; etc. (See Problem Set #3.1, B.)

In this section we turn our attention to methods for solving vector equations. In Section 4.2, another type of vector equation will be introduced. In general, vector equations can be solved in a manner similar to that of scalar equations. Two general methods are illustrated in the following examples.

1. The application of vector operations, such as scalar and vector products, may lead to simplification of an equation.

EXAMPLE 4.1.1   Solve $3\mathbf{x} + 2\mathbf{a} = 5(\mathbf{x} - \mathbf{b})$ for $\mathbf{x}$.

Solution

$$3\mathbf{x} + 2\mathbf{a} = 5\mathbf{x} - 5\mathbf{b}$$
$$(3 - 5)\mathbf{x} = -5\mathbf{b} - 2\mathbf{a}$$
$$-2\mathbf{x} = -(5\mathbf{b} + 2\mathbf{a})$$
$$\mathbf{x} = \frac{5\mathbf{b} + 2\mathbf{a}}{2}.$$

EXAMPLE 4.1.2   Solve $\mathbf{a}x + \mathbf{b}y = \mathbf{c}$ for x, given $\mathbf{a}$, $\mathbf{b}$, and $\mathbf{c}$ coplanar.

Solution

$$\mathbf{b} \times (\mathbf{a}x + \mathbf{b}y) = \mathbf{b} \times \mathbf{c}$$
$$(\mathbf{b} \times \mathbf{a})x + (\mathbf{b} \times \mathbf{b})y = \mathbf{b} \times \mathbf{c}$$
$$(\mathbf{b} \times \mathbf{a}) \cdot [(\mathbf{b} \times \mathbf{a})x] = (\mathbf{b} \times \mathbf{a}) \cdot (\mathbf{b} \times \mathbf{c})$$
$$(\mathbf{b} \times \mathbf{a})^2 x = (\mathbf{b} \times \mathbf{a}) \cdot (\mathbf{b} \times \mathbf{c})$$
$$x = \frac{(\mathbf{b} \times \mathbf{a}) \cdot (\mathbf{b} \times \mathbf{c})}{(\mathbf{b} \times \mathbf{a})^2} \qquad \text{for } \mathbf{b} \times \mathbf{a} \neq \mathbf{0}.$$

EXAMPLE 4.1.3   Solve the system $\mathbf{a} \cdot \mathbf{x} = m$, $\mathbf{b} \times \mathbf{x} = \mathbf{c}$ for $\mathbf{x}$.

Solution

$$\mathbf{a} \times (\mathbf{b} \times \mathbf{x}) = \mathbf{a} \times \mathbf{c}$$
$$(\mathbf{a} \cdot \mathbf{x})\mathbf{b} - (\mathbf{a} \cdot \mathbf{b})\mathbf{x} = \mathbf{a} \times \mathbf{c}$$
$$m\mathbf{b} - (\mathbf{a} \cdot \mathbf{b})\mathbf{x} = \mathbf{a} \times \mathbf{c}$$
$$(\mathbf{a} \cdot \mathbf{b})\mathbf{x} = m\mathbf{b} - (\mathbf{a} \times \mathbf{c})$$
$$\mathbf{x} = \frac{m\mathbf{b} - (\mathbf{a} \times \mathbf{c})}{\mathbf{a} \cdot \mathbf{b}} \text{ providing } \mathbf{a} \cdot \mathbf{b} \neq 0.$$

(Note that $\mathbf{b} \cdot \mathbf{c} = 0$.)

2. The expansion of vectors in component form and the equating of the coefficients of $\mathbf{i}$, $\mathbf{j}$, and $\mathbf{k}$ may lead to explicit solutions of vector equations.

EXAMPLE 4.1.4 Find $\mathbf{x}$ in Example 4.1.3 above if $\mathbf{a} = \mathbf{i} + 2\mathbf{j} - \mathbf{k}$, $m = 3$, $\mathbf{b} = \mathbf{j} - \mathbf{k}$, $\mathbf{c} = \mathbf{j} + \mathbf{k}$.

Solution

By direct substitution in the given equations, we have $x_1 + 2x_2 - x_3 = 3$ and

$$\begin{vmatrix} \mathbf{i} & \mathbf{j} & \mathbf{k} \\ 0 & 1 & -1 \\ x_1 & x_2 & x_3 \end{vmatrix} = (x_3 + x_2)\mathbf{i} - x_1\mathbf{j} - x_1\mathbf{k} = \mathbf{j} + \mathbf{k}.$$

Thus $x_3 + x_2 = 0$, $x_1 = -1$. Also, $2x_2 + x_2 = 4$ so that $x_2 = \frac{4}{3}$, $x_3 = -\frac{4}{3}$. Finally $\mathbf{x} = -\mathbf{i} + \frac{4}{3}\mathbf{j} - \frac{4}{3}\mathbf{k}$.

Check by using the result obtained in Example 4.1.3 to arrive at $\mathbf{x}$.

EXAMPLE 4.1.5 Solve the system $\mathbf{a} \cdot \mathbf{x} = 3$, $\mathbf{b} \cdot \mathbf{x} = 1$, $\mathbf{c} \cdot \mathbf{x} = 0$ for $\mathbf{x}$, given that $\mathbf{a} = \mathbf{i} - 2\mathbf{j} + 3\mathbf{k}$, $\mathbf{b} = 2\mathbf{i} + 3\mathbf{j} + \mathbf{k}$, and $\mathbf{c} = \mathbf{i} + \mathbf{j} + \mathbf{k}$.

Solution

$$x_1 - 2x_2 + 3x_3 = 3$$
$$2x_1 + 3x_2 + x_3 = 1$$
$$x_1 + x_2 + x_3 = 0$$

so that $x_1 = 11$, $x_2 = -5$, $x_3 = -6$ and $\mathbf{x} = 11\mathbf{i} - 5\mathbf{j} - 6\mathbf{k}$.

An important problem that occurs in geometry and physics is the need to express a vector $\mathbf{r}$ in terms of three noncoplanar vectors $\mathbf{a}$, $\mathbf{b}$, and $\mathbf{c}$, that is, if $\mathbf{a} \cdot \mathbf{b} \times \mathbf{c} \neq 0$, then we wish to find scalars α, β, and γ so that $\mathbf{r} = \alpha\mathbf{a} + \beta\mathbf{b} + \gamma\mathbf{c}$.

The scalars α, β, and γ can be expressed in terms of $\mathbf{r}$, $\mathbf{a}$, $\mathbf{b}$, and $\mathbf{c}$ as follows:

Let $\mathbf{r} = \alpha\mathbf{a} + \beta\mathbf{b} + \gamma\mathbf{c}$, then

$$\mathbf{r} \cdot (\mathbf{b} \times \mathbf{c}) = (\alpha\mathbf{a} + \beta\mathbf{b} + \gamma\mathbf{c}) \cdot (\mathbf{b} \times \mathbf{c})$$
$$\mathbf{r} \cdot (\mathbf{b} \times \mathbf{c}) = \alpha\mathbf{a} \cdot (\mathbf{b} \times \mathbf{c}) + \beta\mathbf{b} \cdot (\mathbf{b} \times \mathbf{c}) + \gamma\mathbf{c} \cdot (\mathbf{b} \times \mathbf{c})$$
$$\mathbf{r} \cdot (\mathbf{b} \times \mathbf{c}) = \alpha(\mathbf{a} \cdot \mathbf{b} \times \mathbf{c}) + \beta(\mathbf{b} \cdot \mathbf{b} \times \mathbf{c}) + \gamma(\mathbf{c} \cdot \mathbf{b} \times \mathbf{c})$$
$$\mathbf{r} \cdot (\mathbf{b} \times \mathbf{c}) = \alpha(\mathbf{a} \cdot \mathbf{b} \times \mathbf{c})$$
$$\alpha = \frac{\mathbf{r} \cdot \mathbf{b} \times \mathbf{c}}{\mathbf{a} \cdot \mathbf{b} \times \mathbf{c}}.$$

In a similar manner we obtain

$$\beta = \frac{\mathbf{r} \cdot \mathbf{c} \times \mathbf{a}}{\mathbf{a} \cdot \mathbf{b} \times \mathbf{c}} \quad \text{and} \quad \gamma = \frac{\mathbf{r} \cdot \mathbf{a} \times \mathbf{b}}{\mathbf{a} \cdot \mathbf{b} \times \mathbf{c}}$$

so that we can write

$$\mathbf{r} = \left(\frac{\mathbf{r} \cdot \mathbf{b} \times \mathbf{c}}{\mathbf{a} \cdot \mathbf{b} \times \mathbf{c}}\right)\mathbf{a} + \left(\frac{\mathbf{r} \cdot \mathbf{c} \times \mathbf{a}}{\mathbf{a} \cdot \mathbf{b} \times \mathbf{c}}\right)\mathbf{b} + \left(\frac{\mathbf{r} \cdot \mathbf{a} \times \mathbf{b}}{\mathbf{a} \cdot \mathbf{b} \times \mathbf{c}}\right)\mathbf{c}$$

$$\text{or} \quad \mathbf{r} = \left(\mathbf{r} \cdot \frac{\mathbf{b} \times \mathbf{c}}{\mathbf{a} \cdot \mathbf{b} \times \mathbf{c}}\right)\mathbf{a} + \left(\mathbf{r} \cdot \frac{\mathbf{c} \times \mathbf{a}}{\mathbf{a} \cdot \mathbf{b} \times \mathbf{c}}\right)\mathbf{b} + \left(\mathbf{r} \cdot \frac{\mathbf{a} \times \mathbf{b}}{\mathbf{a} \cdot \mathbf{b} \times \mathbf{c}}\right)\mathbf{c}.$$

If $\mathbf{a} \cdot \mathbf{b} \times \mathbf{c} \neq 0$, the vectors

$$\frac{\mathbf{b} \times \mathbf{c}}{\mathbf{a} \cdot \mathbf{b} \times \mathbf{c}}, \quad \frac{\mathbf{c} \times \mathbf{a}}{\mathbf{a} \cdot \mathbf{b} \times \mathbf{c}}, \quad \frac{\mathbf{a} \times \mathbf{b}}{\mathbf{a} \cdot \mathbf{b} \times \mathbf{c}}$$

are perpendicular to the planes of $\mathbf{b}$ and $\mathbf{c}$, $\mathbf{c}$ and $\mathbf{a}$, $\mathbf{a}$ and $\mathbf{b}$, respectively.

DEFINITION 4.1.1 *If* $\mathbf{a} \cdot \mathbf{b} \times \mathbf{c} \neq 0$, *the system of three vectors*

$$\mathbf{A} = \frac{\mathbf{b} \times \mathbf{c}}{\mathbf{a} \cdot \mathbf{b} \times \mathbf{c}}, \quad \mathbf{B} = \frac{\mathbf{c} \times \mathbf{a}}{\mathbf{a} \cdot \mathbf{b} \times \mathbf{c}}, \quad \mathbf{C} = \frac{\mathbf{a} \times \mathbf{b}}{\mathbf{a} \cdot \mathbf{b} \times \mathbf{c}}$$

is called the **reciprocal system** *to the vectors* $\mathbf{a}$, $\mathbf{b}$, *and* $\mathbf{c}$.

THEOREM 4.1.1 If $\mathbf{a} \cdot \mathbf{b} \times \mathbf{c} \neq 0$ and $\mathbf{r}$ is any vector, then

$$\mathbf{r} = (\mathbf{r} \cdot \mathbf{A})\mathbf{a} + (\mathbf{r} \cdot \mathbf{B})\mathbf{b} + (\mathbf{r} \cdot \mathbf{C})\mathbf{c}$$

where $\mathbf{A}$, $\mathbf{B}$, $\mathbf{C}$ is the reciprocal system to $\mathbf{a}$, $\mathbf{b}$, and $\mathbf{c}$.

In Theorem 4.1.1, the coefficients $\mathbf{r} \cdot \mathbf{A}$, $\mathbf{r} \cdot \mathbf{B}$, and $\mathbf{r} \cdot \mathbf{C}$ are of special interest. In general, a scalar equation of the first degree in a vector $\mathbf{r}$ can be reduced to the form $\mathbf{r} \cdot \mathbf{A} = \alpha$, where $\mathbf{A}$ and α are known. By the definition of the scalar product we have $\mathbf{r} \cdot \mathbf{A} = A \text{ comp}_A \mathbf{r} = \alpha$ or $\text{comp}_A \mathbf{r} = \alpha/A$. Geometrically this states that the terminal point of $\mathbf{r}$ must be on the surface of a plane perpendicular to $\mathbf{A}$ when the initial points of $\mathbf{A}$ and $\mathbf{r}$ coincide (see Figure 4.1.1).

By Theorem 4.1.1 it is clear that three scalar equations

$$\mathbf{r} \cdot \mathbf{A} = \alpha, \quad \mathbf{r} \cdot \mathbf{B} = \beta, \quad \mathbf{r} \cdot \mathbf{C} = \gamma$$

have a unique solution:

$$\mathbf{r} = \alpha\mathbf{a} + \beta\mathbf{b} + \gamma\mathbf{c}$$

where $\mathbf{A}$, $\mathbf{B}$, $\mathbf{C}$ is the reciprocal system to $\mathbf{a}$, $\mathbf{b}$, $\mathbf{c}$.

Geometrically the terminus of $\mathbf{r}$ must be the common point of intersection of the three planes determined by the three scalar equations.

THEOREM 4.1.2 If $\mathbf{A}$, $\mathbf{B}$, $\mathbf{C}$ is the system reciprocal to $\mathbf{a}$, $\mathbf{b}$, $\mathbf{c}$, then $\mathbf{a}$, $\mathbf{b}$, $\mathbf{c}$ will be the reciprocal system to $\mathbf{A}$, $\mathbf{B}$, $\mathbf{C}$.

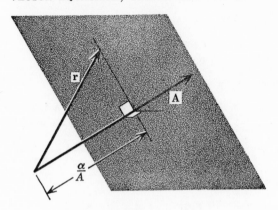

FIGURE 4.1.1 $\mathbf{r} \cdot \mathbf{A} = \alpha$.

THEOREM 4.1.3   If $\mathbf{A}$, $\mathbf{B}$, $\mathbf{C}$ is the system reciprocal to $\mathbf{a}$, $\mathbf{b}$, $\mathbf{c}$, then

$$\mathbf{A} \cdot \mathbf{a} = \mathbf{B} \cdot \mathbf{b} = \mathbf{C} \cdot \mathbf{c} = 1 \text{ and } \mathbf{A} \cdot \mathbf{b} = \mathbf{A} \cdot \mathbf{c} = \mathbf{B} \cdot \mathbf{a} = \mathbf{B} \cdot \mathbf{c} = \mathbf{C} \cdot \mathbf{a} = \mathbf{C} \cdot \mathbf{b} = 0.$$

THEOREM 4.1.4   If $\mathbf{a} \cdot \mathbf{b} \times \mathbf{c} \neq 0$ and $\mathbf{r}$ is any vector, then

$$\mathbf{r} = (\mathbf{r} \cdot \mathbf{a})\mathbf{A} + (\mathbf{r} \cdot \mathbf{b})\mathbf{B} + (\mathbf{r} \cdot \mathbf{c})\mathbf{C}$$

where $\mathbf{A}$, $\mathbf{B}$, $\mathbf{C}$ is the reciprocal system to $\mathbf{a}$, $\mathbf{b}$, and $\mathbf{c}$.

THEOREM 4.1.5   If $\mathbf{A}$, $\mathbf{B}$, $\mathbf{C}$ is the system reciprocal to $\mathbf{a}$, $\mathbf{b}$, $\mathbf{c}$, then $(\mathbf{A} \cdot \mathbf{B} \times \mathbf{C})(\mathbf{a} \cdot \mathbf{b} \times \mathbf{c}) = 1$.

THEOREM 4.1.6   The system of three unit vectors $\mathbf{i}$, $\mathbf{j}$, $\mathbf{k}$ is its own reciprocal system.

We leave Theorems 4.1.2 through 4.1.6 for the reader to prove. The method is to apply Definition 4.1.1, Theorem 4.1.1, and previously discussed procedures.

EXAMPLE 4.1.6   Solve $\mathbf{a}(\mathbf{m} \cdot \mathbf{r}) + \mathbf{b}(\mathbf{p} \cdot \mathbf{r}) + \mathbf{c}(\mathbf{q} \cdot \mathbf{r}) = \mathbf{d}$ for $\mathbf{r}$.

Solution

Assume $\mathbf{a} \cdot \mathbf{b} \times \mathbf{c} \neq 0$ and let $\mathbf{A}$, $\mathbf{B}$, $\mathbf{C}$ be the reciprocal system to $\mathbf{a}$, $\mathbf{b}$, $\mathbf{c}$. Take the scalar product of the equation successively by $\mathbf{A}$, $\mathbf{B}$, and $\mathbf{C}$ to obtain the three scalar equations:

$$\mathbf{m} \cdot \mathbf{r} = \mathbf{A} \cdot \mathbf{d}, \quad \mathbf{p} \cdot \mathbf{r} = \mathbf{B} \cdot \mathbf{d}, \quad \mathbf{q} \cdot \mathbf{r} = \mathbf{C} \cdot \mathbf{d}.$$

Now, for $\mathbf{m} \cdot \mathbf{p} \times \mathbf{q} \neq 0$ and $\mathbf{M}, \mathbf{P}, \mathbf{Q}$, the system reciprocal to $\mathbf{m}, \mathbf{p}$, and $\mathbf{q}$, we have

$$\mathbf{r} = (\mathbf{r} \cdot \mathbf{m})\mathbf{M} + (\mathbf{r} \cdot \mathbf{p})\mathbf{P} + (\mathbf{r} \cdot \mathbf{q})\mathbf{Q}$$

and $\qquad \mathbf{r} = (\mathbf{A} \cdot \mathbf{d})\mathbf{M} + (\mathbf{B} \cdot \mathbf{d})\mathbf{P} + (\mathbf{C} \cdot \mathbf{d})\mathbf{Q}.$

EXAMPLE 4.1.7 Solve $a(\mathbf{m} \cdot \mathbf{r}) + b(\mathbf{p} \cdot \mathbf{r}) + c(\mathbf{q} \cdot \mathbf{r}) = \mathbf{d}$ for $\mathbf{r}$ given that $a = i + 2j - k$, $b = j + k$, $c = 2i - j$, $d = 3i - 2j + k$, $m = i + k$, $p = 2i + j + k$, $q = j + k$.

Solution

First compute the reciprocal systems

$$\mathbf{A} = \tfrac{1}{7}(i + 2j - 2k), \mathbf{B} = \tfrac{1}{7}(i + 2j + 5k), \mathbf{C} = \tfrac{1}{7}(3i - j + k)$$
$$\mathbf{M} = \tfrac{1}{2}(-2j + 2k), \mathbf{P} = \tfrac{1}{2}(i + j - k), \mathbf{Q} = \tfrac{1}{2}(-i + j + k).$$

Now compute the scalar products and substitute

$$\mathbf{r} = -\tfrac{3}{14}(-2j + 2k) + \tfrac{4}{14}(i + j - k) + \tfrac{12}{14}(-i + j + k)$$
$$\mathbf{r} = \tfrac{1}{7}(-4i + 11j + k).$$

In general, vector equations of the first degree in a vector $\mathbf{r}$ are equations containing terms of the following types:

$$a(\mathbf{m} \cdot \mathbf{r}), \quad n\mathbf{r}, \quad \mathbf{b} \times \mathbf{r}, \quad \mathbf{d}.$$

For example, $\mathbf{p} \times (\mathbf{q} \times \mathbf{r}) + \mathbf{b} \times \mathbf{r} + a(\mathbf{m} \cdot \mathbf{r}) + n\mathbf{r} = \mathbf{d}$.

Equations of this general type can often be simplified by the application of identities:

$$\mathbf{p} \times (\mathbf{q} \times \mathbf{r}) = (\mathbf{p} \cdot \mathbf{r})\mathbf{q} - (\mathbf{p} \cdot \mathbf{q})\mathbf{r}.$$

Further reduction may be possible by applying reciprocal systems. If $\mathbf{m} \cdot \mathbf{p} \times \mathbf{q} \neq 0$, we can write $\mathbf{r} = (\mathbf{m} \cdot \mathbf{r})\mathbf{M} + (\mathbf{p} \cdot \mathbf{r})\mathbf{P} + (\mathbf{q} \cdot \mathbf{r})\mathbf{Q}$

so that $\qquad \mathbf{b} \times \mathbf{r} = \mathbf{b} \times [(\mathbf{m} \cdot \mathbf{r})\mathbf{M} + (\mathbf{p} \cdot \mathbf{r})\mathbf{P} + (\mathbf{q} \cdot \mathbf{r})\mathbf{Q}]$
$$= (\mathbf{b} \times \mathbf{M})(\mathbf{m} \cdot \mathbf{r}) + (\mathbf{b} \times \mathbf{P})(\mathbf{p} \cdot \mathbf{r}) + (\mathbf{b} \times \mathbf{Q})(\mathbf{q} \cdot \mathbf{r})$$

and also $\qquad n\mathbf{r} = n[(\mathbf{m} \cdot \mathbf{r})\mathbf{M} + (\mathbf{p} \cdot \mathbf{r})\mathbf{P} + (\mathbf{q} \cdot \mathbf{r})\mathbf{Q}]$
$$= n\mathbf{M}(\mathbf{m} \cdot \mathbf{r}) + n\mathbf{P}(\mathbf{p} \cdot \mathbf{r}) + n\mathbf{Q}(\mathbf{q} \cdot \mathbf{r})$$
$$(\mathbf{p} \cdot \mathbf{q})\mathbf{r} = (\mathbf{p} \cdot \mathbf{q})\mathbf{M}(\mathbf{m} \cdot \mathbf{r}) + (\mathbf{p} \cdot \mathbf{q})\mathbf{P}(\mathbf{p} \cdot \mathbf{r}) + (\mathbf{p} \cdot \mathbf{q})\mathbf{Q}(\mathbf{q} \cdot \mathbf{r}).$$

Thus $\mathbf{p} \times (\mathbf{q} \times \mathbf{r}) + \mathbf{b} \times \mathbf{r} + a(\mathbf{m} \cdot \mathbf{r}) + n\mathbf{r} = \mathbf{d}$ can be written in the form

$$[a + (\mathbf{b} \times \mathbf{M}) + n\mathbf{M} - (\mathbf{p} \cdot \mathbf{q})\mathbf{M}](\mathbf{m} \cdot \mathbf{r})$$
$$+ [\mathbf{q} + (\mathbf{b} \times \mathbf{P}) + n\mathbf{P} - (\mathbf{p} \cdot \mathbf{q})\mathbf{P}](\mathbf{p} \cdot \mathbf{r})$$
$$+ [(\mathbf{b} \times \mathbf{Q}) + n\mathbf{Q} - (\mathbf{p} \cdot \mathbf{q})\mathbf{Q}](\mathbf{q} \cdot \mathbf{r}) = \mathbf{d}.$$

This latter equation can be solved as shown in Examples 4.1.6 and 4.1.7. Thus by applying identities and Theorem 4.1.2, the most general vector equations of the first degree in a vector $\mathbf{r}$ may be reduced to the form $a(\mathbf{m} \cdot \mathbf{r}) + b(\mathbf{p} \cdot \mathbf{r}) + c(\mathbf{q} \cdot \mathbf{r}) = \mathbf{d}$, which has the solution $\mathbf{r} = (\mathbf{A} \cdot \mathbf{d})\mathbf{M} + (\mathbf{B} \cdot \mathbf{d})\mathbf{P} + (\mathbf{C} \cdot \mathbf{d})\mathbf{Q}$ providing $\mathbf{a} \cdot \mathbf{b} \times \mathbf{c} \neq 0$ and $\mathbf{m} \cdot \mathbf{p} \times \mathbf{q} \neq 0$, where **A**, **B**, **C** is the reciprocal system to **a**, **b**, **c** and **M**, **P**, **Q** is the reciprocal system to **m**, **p**, **q**.

4.2   SPACE CURVES

A knowledge of the properties of space curves and surfaces is very important in advanced studies. Physical notions, such as partical motion along a curve and the quantity of fluid flow through a surface, may be represented by the use of vector equations of space curves and surfaces. A detailed analysis of space curves and surfaces requires the calculus and is usually undertaken in more advanced courses.* The discussion that follows will be limited to the consideration of how curves and surfaces can be represented by vector equations.

We recall that the graph of $x^2 + y^2 = c^2$ in a Cartesian coordinate system results in a circle with radius c and center at the origin. If one thinks of $P(x, y)$ as a variable point and O as a fixed point at the origin, then $|\overrightarrow{OP}| = c$ will be a vector equation of the same circle.

Although the two representations of the circle just noted are useful in many applications, they are not completely satisfactory for all applications. Suppose, for example, that a wheel of radius c is rotating at 5 rps in a counterclockwise direction. Either equation, $x^2 + y^2 = c^2$ or $|\overrightarrow{OP}| = c$ will represent the path of a point on the rim of the wheel if its center is at the origin, but neither equation will tell one where the point is at a given time. A parametric representation of the path is more useful in this situation. If the point on the rim is at $(c, 0)$ when $t = 0$, where t is a measure of time, then the coordinates of the point at any time $t \geq 0$ will be $(c \cos 10\pi t, c \sin 10\pi t)$. The parametric representation of the path given by the pair of equations:

$$(1) \qquad \begin{aligned} x &= c \cos 10\pi t \\ y &= c \sin 10\pi t \end{aligned}, \qquad t \geq 0$$

not only describes the path of the point but also tells where the point is at any specified time.

---

* Courses such as differential geometry; see, for example, Chapter 4 of *Geometry and the Imagination* by Hilbert and Cohen-Vossen, Chelsea, 1952.

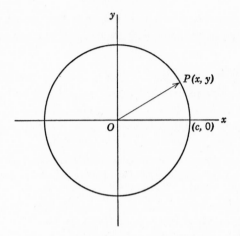

FIGURE 4.2.1 Circle.

The pair of equations (1) can be combined in a vector form:

(2) $\mathbf{r} = (c \cos 10\pi t)\mathbf{i} + (c \sin 10\pi t)\mathbf{j}$

where the initial end of $\mathbf{r}$ is located at the origin and the terminal end at $P(x, y)$. The vector $\mathbf{r}$ in (2) is an example of a bound vector, for its initial point is always at the origin.

DEFINITION 4.2.1 *If $\mathbf{r}$ is a vector with its initial point at the origin and its terminal point at $P(x, y, z)$, then $\mathbf{r} = x\mathbf{i} + y\mathbf{j} + z\mathbf{k}$ is called the* **position vector** *of the point P.*

There are an unlimited number of parametric representations of the circle just considered, hence infinitely many vector equations of the circle. Two additional vector equations of the circle are

(3) $\mathbf{r} = (c \cos \theta)\mathbf{i} + (c \sin \theta)\mathbf{j}, \ 0 \leq \theta \leq 2\pi$

where θ is the positive angle between $\mathbf{r}$ and the positive x-axis.

(4) $\mathbf{r} = \left[c \cos \left(\dfrac{s}{c} \right) \right] \mathbf{i} + \left[c \sin \left(\dfrac{s}{c} \right) \right] \mathbf{j}, \ 0 \leq s \leq 2\pi c$

where s is the length of arc from $(c, 0)$ to $P(x, y)$ on the circle measured in a counterclockwise direction.

In general, a space curve is described by the set of terminal points of a position vector equation

$$\mathbf{r} = X(t)\mathbf{i} + Y(t)\mathbf{j} + Z(t)\mathbf{k}$$

where $x = X(t)$, $y = Y(t)$, $z = Z(t)$ define real valued continuous functions on the parametric interval $[t_0, t_1]$. For curves lying in the xy-coordinate plane, $Z(t)$ is, of course, zero. The parameter t frequently is used to represent time but may also represent other quantities, such as angle or arc length.

EXAMPLE 4.2.1   If a point revolves around an axis at a constant distance, a, from it and at the same time moves parallel to the axis in such a way that its parallel displacement is proportional to the angle of revolution, θ, the resulting curve is called a **circular helix**.

(a) If the z-axis is the axis of the helix and the parameter is the angle of revolution, then the position vector equation of the helix is given by

$$\mathbf{r} = (a \cos \theta)\mathbf{i} + (a \sin \theta)\mathbf{j} + (b\theta)\mathbf{k}$$

where $a > 0$, $b \neq 0$, and θ is the angle of revolution.   Note that

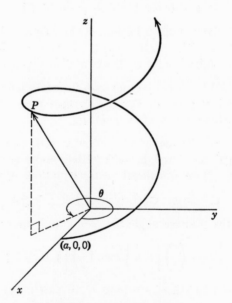

FIGURE 4.2.2   Circular helix.

if b were zero, then the equation would reduce to the equation of the circle considered earlier.

(b) Given that a point on a circular helix starts at $(5, 0, 0)$ when $t = 0$, is at $(5, 0, 2\pi)$ when $t = 1$, and rotates about the z-axis at 3 rps, find the position vector equation of motion of the point with respect to time.

Solution

Since $\theta = 3(2\pi t) = 6\pi t$ and when $t = 1$, we have $b\theta = 2\pi$, $b = \frac{1}{3}$. Thus the position vector equation of the point with respect to time is

$$\mathbf{r} = (5 \cos 6\pi t)\mathbf{i} + (5 \sin 6\pi t)\mathbf{j} + (2\pi t)\mathbf{k}, \qquad t \geq 0.$$

We note that in addition to giving the position of the particle at any time t, the equation can (with the use of calculus) be used to find the velocity and acceleration of the particle at time t.

EXAMPLE 4.2.2 A **cycloid** is a curve that is traced by a point on the circumference of a circle as it rolls along a straight line.

(a) If the straight line is the x-axis and the tracing point (x, y) starts at the origin, then a position vector equation of the cycloid is given by

$$\mathbf{r} = a(\theta - \sin \theta)\mathbf{i} + a(1 - \cos \theta)\mathbf{j}$$

where a is the radius of the circle and θ is the angle through which the circle has revolved.

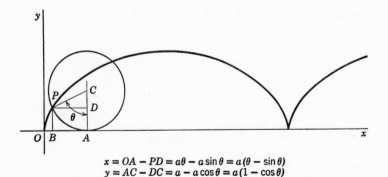

$$x = OA - PD = a\theta - a\sin\theta = a(\theta - \sin\theta)$$
$$y = AC - DC = a - a\cos\theta = a(1 - \cos\theta)$$

FIGURE 4.2.3 Cycloid.

(b) If a circle with radius 5 units rolls along the positive x-axis at 10 rps and the tracing point $P(x, y)$ starts at the origin when $t = 0$, then the position vector equation of motion of the point in terms of time is

$$\mathbf{r} = 5(20\pi t - \sin 20\pi t)\mathbf{i} + 5(1 - \cos 20\pi t)\mathbf{j}, \qquad t \geq 0$$

since $\theta = 10(2\pi t) = 20\pi t$ and $a = 5$.

4.3 SURFACES

A surface, roughly speaking, is a collection of points with a two-dimensional character. (The aggregate of points that constitute a space curve is generally thought of as being one-dimensional in character.) A vector representation of a surface, analogous to that of the space curve given in the preceding section, requires two parameters and is given by a position vector equation of the form:

$$\mathbf{r} = X(u, v)\mathbf{i} + Y(u, v)\mathbf{j} + Z(u, v)\mathbf{k}, \qquad \begin{aligned} u_1 &\leq u \leq u_2, \\ v_1 &\leq v \leq v_2{}^* \end{aligned}$$

EXAMPLE 4.3.1 We show that $\mathbf{r} = (c \sin \phi \cos \theta)\mathbf{i} + (c \sin \phi \sin \theta)\mathbf{j} + (c \cos \phi)\mathbf{k}$, $0 \leq \phi \leq \pi$, $0 \leq \theta \leq 2\pi$ is a position vector representation of a sphere with radius c and center at the origin.

Solution

From Figure 4.3.1 we have $OQ = OP \sin \phi$, $OP = c$,

$$\begin{aligned} x &= OA = OQ \cos \theta = c \sin \phi \cos \theta = X(\phi, \theta) \\ y &= OB = OQ \sin \theta = c \sin \phi \sin \theta = Y(\phi, \theta) \\ z &= OC = OP \cos \phi = c \cos \phi = Z(\phi, \theta) \\ r^2 &= x^2 + y^2 + z^2 = c^2 \end{aligned}$$

EXAMPLE 4.3.2 A surface that is of interest because of its special geometric properties is the **torus** (an inner tube or doughnut are near examples). A vector equation of a torus is

$$\mathbf{r} = [(a + b \cos \alpha) \cos \beta]\mathbf{i} + [(a + b \cos \alpha) \sin \beta]\mathbf{j} + [b \sin \alpha]\mathbf{k}$$
where $0 < b < a$, $0 \leq \alpha \leq 2\pi$, $0 \leq \beta \leq 2\pi$. (Figure 4.3-2.)

---

* This representation applies to most surfaces that occur in practice.

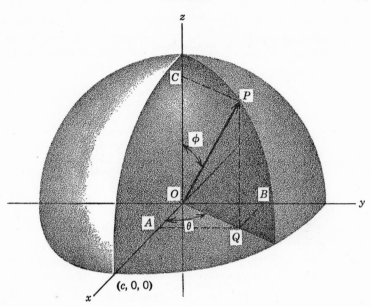

FIGURE 4.3.1   Sphere (first octant).

SUMMARY

4.1   Vector Equations

EXAMPLES 4.1.1 through 4.1.5   Solutions of vector equations.
DEFINITION 4.1.1   **Reciprocal System**
THEOREMS 4.1.1 through 4.1.6   Properties of reciprocal systems.
EXAMPLES 4.1.6 and 4.1.7   Applications of reciprocal systems.

4.2   Space Curves

DEFINITION 4.2.1   *If* **r** *is a vector with its initial point at the origin and its terminal point at* $P(x, y, z)$, *then* $\mathbf{r} = x\mathbf{i} + y\mathbf{j} + z\mathbf{k}$ *is called the position vector of the point* P.
SPACE CURVES   In general, a vector equation of the form

$$\mathbf{r} = X(t)\mathbf{i} + Y(t)\mathbf{j} + Z(t)\mathbf{k}$$

where $x = X(t), y = Y(t), z = Z(t)$ define real valued continuous

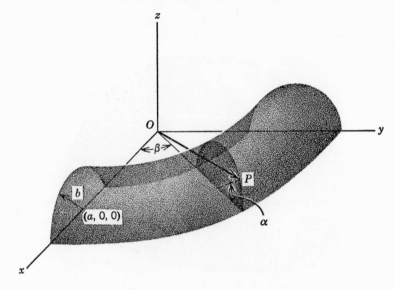

FIGURE 4.3.2   Torus (first octant).

functions on the parametric interval $[t_0, t_1]$ describes a **space curve.**

EXAMPLE 4.2.1   **Circular Helix: r** $= (a \cos \theta)\mathbf{i} + (a \sin \theta)\mathbf{j} + (b\theta)\mathbf{k}$ where $a > 0$, $b \neq 0$, and θ is the angle of revolution.

EXAMPLE 4.2.2   **Cycloid : r** $= a(\theta - \sin \theta)\mathbf{i} + a(1 - \cos \theta)\mathbf{j}$ where a is the radius of the circle and θ is the angle through which the circle has revolved.

4.3   Surfaces

SURFACES   A vector equation of the form

$$\mathbf{r} = X(u, v)\mathbf{i} + Y(u, v)\mathbf{j} + Z(u, v)\mathbf{k}$$

where $x = X(u, v)$, $y = Y(u, v)$, $z = Z(u, v)$ define real valued continuous functions on the parametric interval $[u_0, u_1]$ and $[v_0, v_1]$, describes a **surface.**

EXAMPLE 4.3.1   **Sphere : r** $= (c \sin \phi \cos \theta)\mathbf{i} + (c \sin \phi \sin \theta)\mathbf{j} + (c \cos \phi)\mathbf{k}$ where $0 \leq \phi \leq \pi$, $0 \leq \theta \leq 2\pi$, and c is the radius of the sphere.

EXAMPLE 4.3.2 **Torus:** $\mathbf{r} = [(a + b \cos \alpha) \cos \beta]\mathbf{i} + [(a + b \cos \alpha) \sin \beta]\mathbf{j} + [b \sin \alpha]\mathbf{k}$ where $0 < b < a, 0 \leq \alpha \leq 2\pi, 0 \leq \beta \leq 2\pi$.

PROBLEM SET #4

A. General

1. Solve $2\mathbf{x} - 3\mathbf{a} = \mathbf{b} + 2(\mathbf{x} - 3\mathbf{b})$ for $\mathbf{x}$.

2. Solve $2\mathbf{c} - 3(\mathbf{x} - \mathbf{d}) = 3(\mathbf{c} - \mathbf{d}) - 4\mathbf{x}$ for $\mathbf{x}$.

3. Solve $\mathbf{a}x + \mathbf{b}y = \mathbf{c}$ for y assuming $\mathbf{a}$, $\mathbf{b}$, and $\mathbf{c}$ are coplanar.

4. Solve $2\mathbf{u}x - 3\mathbf{v}y + 2\mathbf{c} = \mathbf{0}$ for x assuming $\mathbf{u}$, $\mathbf{v}$, and $\mathbf{c}$ are coplanar.

5. Show that $\mathbf{a} \cdot \mathbf{b} = 0$ is a necessary and sufficient condition for $\mathbf{a} \times \mathbf{y} = \mathbf{b}$ to have a solution.

6. If $\mathbf{a} \cdot \mathbf{b} = 0, \mathbf{a} \neq \mathbf{0}, \mathbf{b} \neq \mathbf{0}$, then does $\mathbf{a} \times \mathbf{y} = \mathbf{b}$ have a unique solution? Explain.

7. Solve the system: $\mathbf{x} \cdot \mathbf{a} = k, \mathbf{x} \times \mathbf{b} = \mathbf{c}$ for $\mathbf{x}$. (*Note:* $\mathbf{b} \perp \mathbf{c}$.)

8. Let $\mathbf{a} = 2\mathbf{i} - 3\mathbf{j} + \mathbf{k}, \mathbf{b} = -\mathbf{i} + 2\mathbf{j} - \mathbf{k}, \mathbf{c} = \mathbf{i} + 2\mathbf{j} + 3\mathbf{k}$. Solve the system $\mathbf{x} \cdot \mathbf{a} = 10, \mathbf{x} \times \mathbf{b} = \mathbf{c}$ for $\mathbf{x}$ by equating the coefficients of $\mathbf{i}$, $\mathbf{j}$, and $\mathbf{k}$.

9. Given $\mathbf{a} = 2\mathbf{i} + \mathbf{j}, \mathbf{b} = \mathbf{i} - \mathbf{j} + \mathbf{k}, \mathbf{c} = 3\mathbf{j} - \mathbf{k}$, and $\mathbf{d} = -2\mathbf{i} + \mathbf{j} - 3\mathbf{k}$, solve $(\mathbf{a} \times \mathbf{r}) \times \mathbf{b} + (\mathbf{c} \cdot \mathbf{r})\mathbf{d} = 3\mathbf{a}$ for $\mathbf{r}$.

10. Using $\mathbf{a}$, $\mathbf{b}$, $\mathbf{c}$, and $\mathbf{d}$ as given in the preceding problem, solve $2(\mathbf{a} \cdot \mathbf{b} \times \mathbf{r})\mathbf{c} - 3(\mathbf{d} \times \mathbf{r}) = \mathbf{b}$ for $\mathbf{r}$.

11. Solve the linear system

$$x - y + 2z = 3$$
$$2x + y - z = 7$$
$$3y - 2z = 4$$

by letting $\mathbf{a} = \mathbf{i} + 2\mathbf{j}, \mathbf{b} = -\mathbf{i} + \mathbf{j} + 3\mathbf{k}, \mathbf{c} = 2\mathbf{i} - \mathbf{j} - 2\mathbf{k}$, and $\mathbf{d} = 3\mathbf{i} + 7\mathbf{j} + 4\mathbf{k}$. Write $x\mathbf{a} + y\mathbf{b} + z\mathbf{c} = \mathbf{d}$ and take scalar products of both sides with $\mathbf{b} \times \mathbf{c}$, etc.

12. Solve the system $\mathbf{a} \cdot \mathbf{x} = 3, \mathbf{b} \cdot \mathbf{x} = 0, \mathbf{c} \cdot \mathbf{x} = -16$ for $\mathbf{x}$, given that $\mathbf{a} = \mathbf{i} - 3\mathbf{j} + \mathbf{k}, \mathbf{b} = -\mathbf{i} + 2\mathbf{j} - \mathbf{k}, \mathbf{c} = \mathbf{i} - \mathbf{j} - 3\mathbf{k}$.

13. Let $\mathbf{a} = 2\mathbf{i} + 3\mathbf{j} + \mathbf{k}, \mathbf{b} = -\mathbf{i} + 2\mathbf{k}, \mathbf{c} = -\mathbf{i} + \mathbf{j}$, and $\mathbf{r} = \mathbf{i} - 2\mathbf{j} + 2\mathbf{k}$. Find scalars α, β, and γ so that $\mathbf{r} = \alpha\mathbf{a} + \beta\mathbf{b} + \gamma\mathbf{c}$.

14. Let $\mathbf{a} = 12\mathbf{i} - 8\mathbf{j} + 4\mathbf{k}, \mathbf{b} = 4\mathbf{i} + 2\mathbf{j}, \mathbf{c} = 3\mathbf{i} - 2\mathbf{j} + \mathbf{k}$, and $\mathbf{r} = \mathbf{i} - 2\mathbf{j} + 2\mathbf{k}$. Find scalars α, β, and γ so that $\mathbf{r} = \alpha\mathbf{a} + \beta\mathbf{b} + \gamma\mathbf{c}$.

15.* Prove that if $\mathbf{a} \cdot \mathbf{r} = \alpha, \mathbf{b} \cdot \mathbf{r} = \beta, \mathbf{c} \cdot \mathbf{r} = \gamma, \mathbf{d} \cdot \mathbf{r} = \delta$, then $\alpha(\mathbf{b} \cdot \mathbf{c} \times \mathbf{d}) + \beta(\mathbf{c} \cdot \mathbf{a} \times \mathbf{d}) + \gamma(\mathbf{a} \cdot \mathbf{b} \times \mathbf{d}) = \delta(\mathbf{a} \cdot \mathbf{b} \times \mathbf{c})$.

16. Simplify
    (a) $(\mathbf{a} \times \mathbf{b}) \cdot (\mathbf{c} \times \mathbf{r}) + (\mathbf{p} \cdot \mathbf{q} \times \mathbf{r}) - 5 = 0$ to the form $\mathbf{u} \cdot \mathbf{r} = \alpha$.
    (b) $(\mathbf{a} \times \mathbf{r}) \cdot \mathbf{b} - [(\mathbf{c} \times \mathbf{r}) \times \mathbf{d}] \cdot \mathbf{p} + 2 = 0$ to the form $\mathbf{v} \cdot \mathbf{r} = \beta$.

* Optimal problems associated with reciprocal systems.

17.* Find the reciprocal system **A**, **B**, **C** to
   (a) $\mathbf{a} = 2\mathbf{i} - \mathbf{j}$, $\mathbf{b} = \mathbf{j} + \mathbf{k}$, $\mathbf{c} = \mathbf{i} + \mathbf{j} - \mathbf{k}$
   (b) $\mathbf{a} = \mathbf{i} + 2\mathbf{k}$, $\mathbf{b} = 2\mathbf{j}$, $\mathbf{c} = 2\mathbf{i} + \mathbf{j} + 5\mathbf{k}$.

18.* Prove Theorems 4.1.2 through 4.1.4.

19.* Prove Theorems 4.1.5 and 4.1.6.

20.* Prove that if $\mathbf{a} \cdot \mathbf{b} \times \mathbf{c} \neq 0$, then for any scalar product $\mathbf{d} \cdot \mathbf{r}$ we have $\mathbf{d} \cdot \mathbf{r} = (\mathbf{d} \cdot \mathbf{A})(\mathbf{a} \cdot \mathbf{r}) + (\mathbf{d} \cdot \mathbf{B})(\mathbf{b} \cdot \mathbf{r}) + (\mathbf{d} \cdot \mathbf{C})(\mathbf{c} \cdot \mathbf{r})$, where **A**, **B**, **C** is the reciprocal system to **a**, **b**, and **c**.

21.* Using the technique of reciprocal systems, prove that

$$(\mathbf{p} \cdot \mathbf{q} \times \mathbf{r})(\mathbf{a} \cdot \mathbf{b} \times \mathbf{c}) = \begin{vmatrix} \mathbf{p} \cdot \mathbf{a} & \mathbf{p} \cdot \mathbf{b} & \mathbf{p} \cdot \mathbf{c} \\ \mathbf{q} \cdot \mathbf{a} & \mathbf{q} \cdot \mathbf{b} & \mathbf{q} \cdot \mathbf{c} \\ \mathbf{r} \cdot \mathbf{a} & \mathbf{r} \cdot \mathbf{b} & \mathbf{r} \cdot \mathbf{c} \end{vmatrix}.$$

B. Geometric

1. (a) Plot the curve whose vector equation is

   $$\mathbf{r} = \tfrac{1}{2}t\mathbf{i} + t^2\mathbf{j} \qquad \text{for } 0 \leq t \leq 4.$$

   (b) Obtain an equation of the curve in Cartesian coordinates by eliminating the parameter t from the corresponding parametric equations. What restrictions must be placed on x and y for the Cartesian equation to produce the same curve?

2. Plot the curve whose vector equation is

   $$\mathbf{r} = (\tfrac{1}{2}t^2)\mathbf{i} + (\tfrac{1}{4}t^3)\mathbf{j} \qquad \text{for } -\infty < t < \infty.$$

3. Plot the curve whose vector equation is

   $$\mathbf{r} = (4 \cos t)\mathbf{i} + (3 \sin t)\mathbf{j}.$$

4. Plot the curve whose vector equation is

   $$\mathbf{r} = (4 \cos \theta)\mathbf{i} + (3 \sin \theta)\mathbf{j} + (2\theta)\mathbf{k}.$$

5. Plot the curve whose vector equation is

   $$\mathbf{r} = (a \cos \theta)\mathbf{i} + (b \sin \theta)\mathbf{j} + (c\theta)\mathbf{k}; a > 0, b > 0, \text{ and } c > 0.$$

6. Plot the curve whose vector equation is

   $$\mathbf{r} = (2r \cos \theta + r \cos 2\theta)\mathbf{i} + (2r \sin \theta - r \sin 2\theta)\mathbf{j}; r \text{ constant.}$$

7. Let an object move counterclockwise on an elliptical path with center at the origin, major axis 10, minor axis 6, and starting point (5, 0). If the object completes one "orbit" every 3 seconds, write a position vector equation of the motion of the object with respect to the time in seconds.

   * Optional problems associated with reciprocal systems.

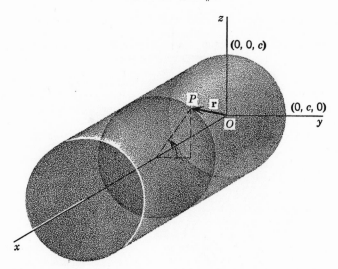

FIGURE #4-1   Circular cylinder.

8. Let an object move clockwise on an elliptical path with center at the origin, major axis 16, minor axis 10, and starting point (8, 0). If the object completes one "orbit" every $\frac{1}{4}$ second, write a position vector equation of the motion of the object with respect to time in seconds.

9. Find a position vector equation in terms of a parameter t of the curve whose equation in rectangular coordinates is given by $y(1 + x^2) = 8$ and such that $y = at^2 + bt + c$ and $y = 8$ when $t = 0$, $y = 4$ when $t = \pm 1$.   Plot the curve.

In Problems 10, 11, and 12 indicate the geometric significance of the parameters u and v in each of the vector equations of surfaces.

10. Circular cylinder: $\mathbf{r} = u\mathbf{i} + (c \cos v)\mathbf{j} + (c \sin v)\mathbf{k}$.

11. Circular cone: $\mathbf{r} = (cu \cos v)\mathbf{i} + (cu \sin v)\mathbf{j} + u\mathbf{k}$.

12. Ellipsoid: $\mathbf{r} = (a \sin u \cos v)\mathbf{i} + (b \sin u \sin v)\mathbf{j} + (c \cos u)\mathbf{k}$ for $0 \leq v \leq 2\pi$, $0 \leq u \leq \pi$.   What is the surface if $a = b = c$?

C. Physics

1. A common family of curves used in the design of gears is called an involute of a circle.   Let a thread be wound around a circle.   The path traced by the free end when the stretched thread is unwound (see Figure #4-4) is called the involute of the circle.

Using Figure #4-4, find a position vector equation for the involute of a circle of radius c using the angle θ as a parameter.

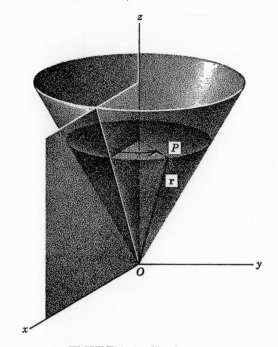

FIGURE #4-2   Circular cone.

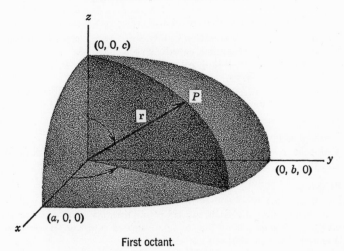

First octant.

FIGURE #4-3   Ellipsoid.

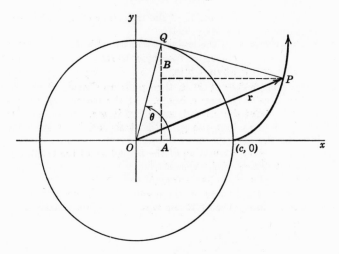

FIGURE #4-4   Involute of a circle.

2. The path of a projectile on the surface of the earth (neglecting air resistance) is given approximately by the vector equation

$$\mathbf{r} = [(v_0 \cos \theta_0)t]\mathbf{i} + [(v_0 \sin \theta_0)t - \tfrac{1}{2}gt^2]\mathbf{j}$$

where v_0 is the initial velocity
  θ_0 is the angle of departure
  t is the time (>0)
  g is the gravitational constant for earth (approximately 32 ft/sec² at the surface).

The direction and speed of the projectile are given by the velocity vector

$$\mathbf{v} = (v_0 \cos \theta_0)\mathbf{i} + (v_0 \sin \theta_0 - gt)\mathbf{j} \qquad \text{for } t > 0.$$

Given that $v_0 = 160$ ft/sec, $\theta_0 = 45°$, $g = 32$ ft/sec²,
  (a) find the horizontal range of the projectile and the time it takes to achieve this range.   Find the velocity vector at this time.
  (b) find the time when the projectile reaches its highest point and find this maximum altitude.   Find the velocity vector at the highest point.
  (c) sketch a graph of the path of the projectile for $y > 0$.

3. A plane in level flight at 800 ft/sec at an altitude of 1,600 ft releases an object.   Select a Cartesian coordinate system so that the plane is at the origin flying in the positive x-axis direction when the object is released (the ground is at $y = -1600$).   Neglect air resistance.

    (a) Write the vector equation of the trajectory of the object and the vector equation of its velocity.

    (b) Sketch the path of the trajectory of the object.

    (c) Find the time it takes for the object to strike the ground and the velocity vector when it hits.

4. A bombing plane flies at 50 ft, then pulls up sharply at an angle of 60° at which time it releases a bomb. At the moment of release the plane is 100 ft above the ground, going 800 ft/sec. Select a Cartesian coordinate system so that the plane is at the origin at the time the bomb is released.

    (a) Write the vector equation of the trajectory of the bomb and the vector equation of the velocity.

    (b) Sketch a plot of the trajectory of the bomb.

    (c) Find the time it takes for the bomb to hit the ground and evaluate the velocity vector at the moment the bomb strikes.

5

vector functions

5.1 VECTOR FUNCTIONS IN A SINGLE INDEPENDENT VARIABLE

The function concept is one of the most important and useful notions in mathematics. The concept of function is introduced to most students in an elementary algebra course and assumes a prominent role in most of their remaining mathematics training. The types of functions first encountered are concerned with relationships between real numbers, that is, with sets of ordered pairs of real numbers. There are, however, many other types of functions that are formed from sets of ordered pairs of elements other than real numbers. A vector function is an example.

The reader will recall that if x represents an element in one set and y represents an element in another set and if there is a correspondence such that for each x we have a unique y, then we write $y = f(x)$ and say that a function is defined. The set of ordered pairs (x, y) that make the condition $y = f(x)$ true is often said to be the function. The set of first elements x in the set of ordered pairs (x, y) is called the **domain** of the function; the set of second elements y is called the **range** of the function. Any symbol, such as x, used to denote an arbitrary element in the domain of a function is called an **independent variable**; any symbol used to denote an arbitrary element in the range of a function is called a **dependent variable.** If the domain and range of a function consist of real numbers, the function is said to be a **real valued function** of a real variable. If the domain consists of ordered pairs of real numbers and the range consists of real numbers, the function is said to be a real valued function in two real variables. In the following material we are concerned with functions whose domains consist of a single set of real numbers and whose ranges consist of vectors.

DEFINITION 5.1.1 *A* **Vector Function** (*in one independent real variable*) *is a function having a set of real numbers as its domain and a set of vectors as its range.*

Vector functions are frequently specified by vector equations. Many of the equations encountered in Chapter 4 can be considered as defining vector functions, for example, $\mathbf{r} = \overrightarrow{f(\theta)}$, where

(i) $f = \{(\theta, \mathbf{r})/\mathbf{r} = (k \cos \theta)\mathbf{i} + (k \sin \theta)\mathbf{j}, \qquad 0 \leq \theta \leq 2\pi\}$*

(ii) $g = \{(t, \mathbf{r})/\mathbf{r} = (5 \cos 6\pi t)\mathbf{i} + (5 \sin 6\pi t)\mathbf{j} + (2\pi t)\mathbf{k}, \qquad t \geq 0\}$

(iii) $F = \{(t, \mathbf{r})/\mathbf{r} = [(v_0 \cos \theta_0)t]\mathbf{i} + [(v_0 \sin \theta_0)t - \frac{1}{2}gt^2]\mathbf{j}, \; t \geq 0\}$

(iv) $G = \{(t, \mathbf{v})/\mathbf{v} = (v_0 \cos \theta_0)\mathbf{i} + [(v_0 \sin \theta_0) - gt]\mathbf{j}, \qquad t \geq 0\}$

We first consider how to find the range value of a vector function, given a domain value of the function, that is, let us evaluate:

(a) $\overrightarrow{f\left(\dfrac{\pi}{6}\right)}$ (b) $\overrightarrow{g\left(\dfrac{1}{6}\right)}$ (c) $\overrightarrow{F(2)}$ (d) $\overrightarrow{G(2)}$

where the functions f, g, F, and G are as defined above.

* Read: "The vector function f is the set of all ordered pairs $(\theta, \mathbf{r})$ such that $\mathbf{r} = (k \cos \theta)\mathbf{i} + (k \sin \theta)\mathbf{j}, 0 \leq \theta \leq 2\pi$." $\mathbf{r} = \overrightarrow{f(\theta)}$ defines a vector $\mathbf{r}$ in the range of f corresponding to θ in the domain of f.

Solutions

(a) $\overrightarrow{f\left(\frac{\pi}{6}\right)} = \left(k \cos \frac{\pi}{6}\right)\mathbf{i} + \left(k \sin \frac{\pi}{6}\right)\mathbf{j} = \left(\frac{\sqrt{3}k}{2}\right)\mathbf{i} + \left(\frac{k}{2}\right)\mathbf{j}$

(b) $\overrightarrow{g\left(\frac{1}{6}\right)} = 5 \cos(\pi)\mathbf{i} + 5 \sin(\pi)\mathbf{j} + \frac{\pi}{3}\mathbf{k} = -5\mathbf{i} + \frac{\pi}{3}\mathbf{k}$

(c) $\overrightarrow{F(2)} = [(v_0 \cos \theta_0)2]\mathbf{i} + [(v_0 \sin \theta_0)2 - 2g]\mathbf{j}$

(d) $\overrightarrow{G(2)} = (v_0 \cos \theta_0)\mathbf{i} + (v_0 \sin \theta_0 - 2g)\mathbf{j}$

The geometric and physical interpretations of vector functions depend on the context in which they appear. For example, if the vectors in the range of the functions f, g, and F are thought of as position vectors (i.e., as vectors with their initial ends fixed at the origin), then the terminal ends of the vectors will trace well-known space curves as the independent variables assume their specified values. The functions f and g will trace a circle and a helix, respectively (see Section 4.2), while F will trace the flight of a ballistic projectile in a vacuum (the trace will be a parabola) with initial velocity v_0 and an initial angle of elevation θ_0 (see Problem Set #4, C-2).

The vectors in the range of G, on the other hand, can be interpreted as the velocity of the projectile (at time t) whose path is

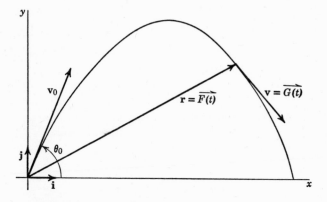

FIGURE 5.1.1   Trajectory and velocity of a projectile.

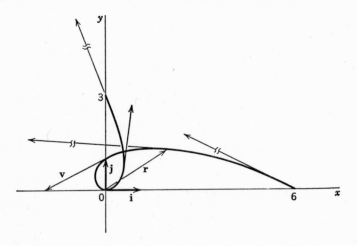

FIGURE 5.1.2 $\mathbf{r} = t(1 - t^2)\mathbf{i} - t^2(t + 2)\mathbf{j}$.

given by the function F, that is, for each t_0 we can obtain a position vector $\overrightarrow{F(t_0)}$ at whose terminal point we can construct the vector $\overrightarrow{G(t_0)}$.   Associated with each point on the flight path of the projectile is a velocity vector that indicates the direction and magnitude of the velocity of the projectile when it is at that point. Note that t does not appear on the graph.

As a further example, let

$$\mathbf{r} = t(1 - t^2)\mathbf{i} + t^2(t + 2)\mathbf{j} \qquad \text{for } -2 \le t \le 1$$

be the vector function of position of an object.   It can also be shown (with the calculus) that

$$\mathbf{v} = (1 - 3t^2)\mathbf{i} + t(3t + 4)\mathbf{j}$$

will describe the velocity of the object at time t.

We first evaluate the vector functions for convenient values of t:

t	$\mathbf{r}$	$\mathbf{v}$
-2	$6\mathbf{i}$	$-11\mathbf{i} + 4\mathbf{j}$
-1	$\mathbf{j}$	$-2\mathbf{i} - \mathbf{j}$
0	$\mathbf{0}$	$\mathbf{i}$
1	$3\mathbf{j}$	$-2\mathbf{i} + 7\mathbf{j}$

Graphically, the range values of **r** can be plotted to describe the path of the object. At the terminal ends of **r** we can plot the corresponding values of **v**. Note that the domain values t do not appear directly on the graph, rather, the graph represents the trace of the terminal points of the range vectors **r** for the values $-2 \le t \le 1$.

5.2 VECTOR AND SCALAR POINT FUNCTIONS

We have already noted that it is not necessary that the domains of functions be sets of real numbers. In many applications it is convenient to consider functions whose domains are sets of points in a specified region. Functions whose domains consist of points in a given region are commonly called "Point Functions."

DEFINITION 5.2.1 *A* **scalar point function** *is a function with a set of points as its domain and a set of real numbers as its range.*

DEFINITION 5.2.2 *A* **vector point function** *is a function with a set of points as its domain and a set of vectors as its range.*

Scalar point functions and vector point functions are frequently referred to as **scalar fields** and **vector fields**. Scalar fields are often associated with problems dealing with temperature, density, pressure, potential, and so on. Vector fields are associated with problems involving gravitational forces, electrical forces, heat and fluid flow, and so forth. In many problems vector and scalar fields are closely associated and are used together to examine different aspects of the problems.

To illustrate the evaluation and plotting of scalar and vector fields, let us consider a domain defined by

$$R = \{p/(x-2)^2 \le 2y \text{ and } y \le 2, \text{ where } (x, y) \text{ are the coordinates of point } p \text{ in a Cartesian system.}\}$$

Let a scalar and a vector point function be defined by

$$f = \left\{ (p, t) \Big/ t = \frac{x^2 - 4x + 7}{3y + 1}, \text{ where } p(x, y) \text{ is in } R \right\}$$

$$G = \left\{ (p, \mathbf{d}) \Big/ \mathbf{d} = \left(\frac{2x - 4}{3y + 1}\right)\mathbf{i} - \left(\frac{3(x^2 - 4x + 7)}{(3y + 1)^2}\right)\mathbf{j}, \text{ where } p(x, y) \text{ is in } R. \right\}$$

TABLE 5.2.1

$p(x, y)$		t	$\mathbf{d}$
0	2	1	$-\frac{4}{7}\mathbf{i} - \frac{3}{7}\mathbf{j}$
1	1	1	$-\frac{1}{2}\mathbf{i} - \frac{3}{4}\mathbf{j}$
1	2	$\frac{4}{7}$	$-\frac{2}{7}\mathbf{i} - \frac{12}{49}\mathbf{j}$
2	0	3	$-9\mathbf{j}$
2	1	$\frac{3}{4}$	$-\frac{9}{16}\mathbf{j}$
2	2	$\frac{3}{7}$	$-\frac{9}{49}\mathbf{j}$
3	1	1	$\frac{1}{2}\mathbf{i} - \frac{3}{4}\mathbf{j}$
3	2	$\frac{4}{7}$	$\frac{2}{7}\mathbf{i} - \frac{12}{49}\mathbf{j}$
4	2	1	$\frac{4}{7}\mathbf{i} - \frac{3}{7}\mathbf{j}$

The scalar point function f can be graphically illustrated by evaluating the range values of f for convenient values of p and by "flagging" these values at the corresponding domain points in R (see Figure 5.2.2). The vector point function G can also be illustrated as shown in Figure 5.2.3. The initial points of the range vectors of G are placed at the corresponding domain points of G.

In a scalar field a real number is associated with each point in the field; in a vector field, on the other hand, a vector is associated with each point in the field. A few additional examples

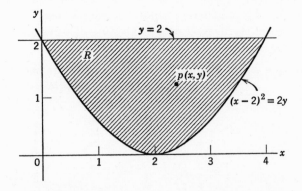

FIGURE 5.2.1 $R = \{p/(x - 2)^2 \leq 2y \text{ and } y \leq 2\}$.

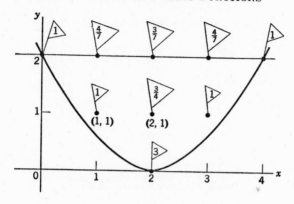

FIGURE 5.2.2 $f = \left\{ (p,t)/t = \dfrac{x^2 - 4x + 7}{3y + 1} \right\}.$

are given below to help illustrate the applications of the concepts just described.

EXAMPLE 5.2.1 The pressure at any point in a fluid (gas or a liquid) can be described by a scalar point function. In particular, the pressure

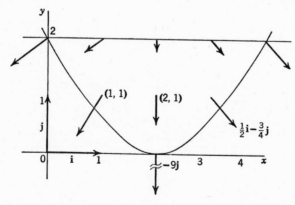

FIGURE 5.2.3 $G = \left\{ (p,\mathbf{d})/\mathbf{d} = \left(\dfrac{2x - 4}{3y + 1} \right)\mathbf{i} + \left(\dfrac{3(x^2 - 4x + 7)}{(3y + 1)^2} \right)\mathbf{j} \right\}$

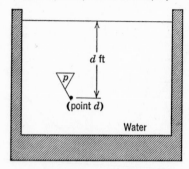

FIGURE 5.2.4   A scalar field defined by the pressure at a point.

at a point within an open container of water is given by the following scalar point function.   (See Figure 5.2.4.)

$$P = \{(d,p)/p = (2120 + 62.5d),$$ where d is a point which is d units beneath the surface measured in feet.$\}$

EXAMPLE 5.2.2   A thin rectangular plate 10 cm wide with a semi-infinite length is located in a Cartesian coordinate system as indicated in Figure 5.2.5.   The surfaces are insulated, the long edges are kept at 0° C and the short edge is kept at 10° C.   The temperature at a point in the plate tends to zero as the distance between the point and the edge

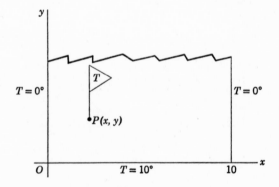

FIGURE 5.2.5   A scalar field defined by the temperature distribution in a plate.

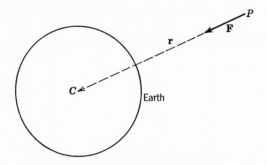

FIGURE 5.2.6   A vector field defined by the gravitational field of the earth.

$y = 0$ tends to infinity.   The steady-state temperature at any point P within the plate is given by the following scalar point function:

$$U = \left\{ (P, T)/T = \frac{20}{\pi} \arctan \left(\frac{\sin (\pi x/10)}{\sinh (\pi y/10)} \right), \text{ where } P \text{ is} \right.$$

$$\left. \text{a point with coordinates } (x, y). \right\}^{*}$$

We write $T = U(P)$ to represent the temperature at the point P.

EXAMPLE 5.2.3   The gravitational field of the earth can be described by a vector point function:

$$H = \left\{ (P, \mathbf{F}) \middle/ \mathbf{F} = \left(\frac{Gm_e}{r^3} \right) \mathbf{r}, \text{ where } \mathbf{r} \text{ is a vector from the point } P \right.$$

$$\left. \text{to the center of the earth.} \right\}$$

Note: $G = 6.670(10^{-8})$ dyne cm^2/gm^2 (the gravitational constant)
   $m_e = 5.98(10^{27})$ gm (the mass of the earth)
   $r = |\mathbf{r}|$ (the distance to the center of the earth)

   $\mathbf{F} = \overrightarrow{H(P)}$ is then the force per unit mass (or gravitational intensity) on a particle at P and is directed toward the center of the earth.

To illustrate further, when P is $16.05(10^8)$ cm (approximately 10,000

* This scalar point function gives a good approximation of the temperature distribution if the semi-infinite length is replaced with a finite length that is large relative to the width and the resulting top edge is kept at 0° C.

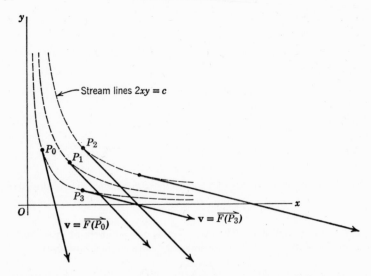

FIGURE 5.2.7   A vector field associated with a fluid flow around a corner.

miles) from the earth's center, we have

$$\mathbf{F} = \frac{(6.67 \times 10^{-8})(5.98 \times 10^{27})}{(16.05 \times 10^{8})^{3}} \overrightarrow{PC} = (9.63 \times 10^{-8})\overrightarrow{PC}$$

$$|\mathbf{F}| = |(9.63 \times 10^{-8})\overrightarrow{PC}| = (9.63 \times 10^{-8})(16.05 \times 10^{8}) = 153 \text{ dyne/gm}$$

EXAMPLE 5.2.4 The velocity vectors associated with a uniform fluid flow around a corner of a noncompressible fluid, such as water, is given by the following vector point function:

$$F = \{(p, \mathbf{v})/\mathbf{v} = 2x\mathbf{i} - 2y\mathbf{j}, \text{ where } p \text{ is the point}$$
$$\text{with coordinates } (x, y).\}$$

We often write $\mathbf{v} = \overrightarrow{F(p)}$ for the velocity of a particle of fluid at the point p.

For $p(x, y)$ we have $\overrightarrow{F(p)}$:

$$p_0(\tfrac{1}{2}, 2) \qquad\qquad \overrightarrow{F(p_0)} = \mathbf{i} - 4\mathbf{j}$$
$$p_1(\tfrac{3}{2}, \tfrac{3}{2}) \qquad\qquad \overrightarrow{F(p_1)} = 3\mathbf{i} - 3\mathbf{j}$$
$$p_2(2, 2) \qquad\qquad \overrightarrow{F(p_2)} = 4\mathbf{i} - 4\mathbf{j}$$
$$p_3(2, \tfrac{1}{2}) \qquad\qquad \overrightarrow{F(p_3)} = 4\mathbf{i} - \mathbf{j}$$

SUMMARY

5.1   Vector Functions in a Single Independent Variable

DEFINITION 5.1.1   *A* **Vector Function** (*in one independent real variable*) *is a function having a set of real numbers as its domain and a set of vectors as its range.*

EXAMPLES of Vector Functions:

(i) $f = \{(\theta, \mathbf{r})/\mathbf{r} = (k \cos \theta)\mathbf{i} + (k \sin \theta)\mathbf{j}, \qquad 0 \leq \theta \leq 2\pi\}$

(ii) $g = \{(t, \mathbf{r})/\mathbf{r} = (5 \cos 6\pi t)\mathbf{i} + (5 \sin 6\pi t)\mathbf{j} + (2\pi t)\mathbf{k}, \quad t \geq 0\}$

(iii) $F = \{(t, \mathbf{r})/\mathbf{r} = [(v_0 \cos \theta_0)t]\mathbf{i} + [(v_0 \sin \theta_0)t - \frac{1}{2}gt^2]\mathbf{j}, \; t \geq 0\}$

(iv) $G = \{(t, y)/\mathbf{v} = (v_0 \cos \theta_0)\mathbf{i} + [(v_0 \sin \theta_0) - gt]\mathbf{j}, \qquad t \geq 0\}$

5.2   Vector and Scalar Point Functions

DEFINITION 5.2.1   *A* **scalar point function** *is a function with a set of points as its domain and a set of real numbers as its range.*

DEFINITION 5.2.2   *A* **vector point function** *is a function with a set of points as its domain and a set of vectors as its range.*

EXAMPLES 5.2.1   (Fluid pressure), 5.2.2 (Temperature), 5.2.3 (Gravitational field), 5.2.4 (Fluid flow).

PROBLEM SET #5

A. and B.   General and Geometric

1. For the examples of vector functions immediately following Definition 5.1.1, evaluate

$$\text{(a)} \; \overrightarrow{f\left(\frac{\pi}{3}\right)} \cdot \overrightarrow{g\left(\frac{1}{12}\right)} \qquad \text{(b)} \; \overrightarrow{g\left(\frac{1}{18}\right)} \times \overrightarrow{F(3)}$$

2. For the examples of vector functions immediately following Definition 5.1.1, evaluate $\overrightarrow{g(\frac{1}{18})} \cdot \overrightarrow{g(\frac{1}{6})} \times \overrightarrow{g(0)}$.

3. Let the vector function f be defined as follows:

$$f = \{(t, \mathbf{r})/\mathbf{r} = (4 \cos t)\mathbf{i} + (2 \sin t)\mathbf{j}, 0 \leq t < 2\pi\}$$

(a) Find $\overrightarrow{f(0)}, \overrightarrow{f\left(\frac{\pi}{6}\right)}, \overrightarrow{f\left(\frac{\pi}{3}\right)}, \overrightarrow{f\left(\frac{\pi}{2}\right)}.$

(b) Sketch a graph of **r** interpreted as a position-vector.

(c) Draw in the vectors $\overrightarrow{f\left(\frac{\pi}{6}\right)}$, $\overrightarrow{f\left(\frac{\pi}{3}\right)}$, and $\left[\overrightarrow{f\left(\frac{\pi}{3}\right)} - \overrightarrow{f\left(\frac{\pi}{6}\right)}\right]$ in the graph of part (b).

4. Let the vector function g be defined as follows:

$$g = \{(t, \mathbf{r})/\mathbf{r} = t\mathbf{i} + t^2\mathbf{j}, \quad t \geq 0\}$$

(a) Find $\overrightarrow{g(0)}$, $\overrightarrow{g(1)}$, $\overrightarrow{g(2)}$, $\overrightarrow{g(3)}$, and $\overrightarrow{g(4)}$.

(b) Sketch a graph of **r** interpreted as a position vector.

(c) Locate and draw in the vectors $\overrightarrow{g(1)}$, $\overrightarrow{g(2)}$, and $[\overrightarrow{g(2)} - \overrightarrow{g(1)}]$ in the graph of part (b).

5. Graph $\mathbf{r} = \overrightarrow{g(t)}$ of Problem 4 above again.

(a) Locate and graph the vector $\dfrac{\overrightarrow{g(\frac{3}{2})} - \overrightarrow{g(1)}}{\frac{1}{2}}$

(b) Locate and graph the vector $\dfrac{\overrightarrow{g(1.1)} - \overrightarrow{g(1)}}{0.1}$

6. Graph $\mathbf{r} = \overrightarrow{g(t)}$ of Problem 4 again.

(a) Let h be a small positive number and locate the vector

$$\frac{\overrightarrow{g(1 + h)} - \overrightarrow{g(1)}}{h}.$$

(b) What constant vector does the vector of part (a) approach as h is made to approach zero? Graph this constant vector so that its initial end is on the terminal end of $\overrightarrow{g(1)}$.

(c) What is the geometric relationship of the constant vector to the curve?

7. In problem 6, replace $\dfrac{\overrightarrow{g(1 + h)} - \overrightarrow{g(1)}}{h}$ with $\dfrac{\overrightarrow{g(2 + h)} - \overrightarrow{g(2)}}{h}$ and repeat.

C. Physics

1. Using the function defined in Example 5.2.1, find $p = P(5280)$, that is, the pressure at a point 5280 ft below the surface of the waters of a very deep lake.

2. If the initial velocity of a projectile is $v_0 = 1500$ ft/sec and its initial angle of elevation is $\theta_0 = \pi/3$, graph the position vector function of the projectile. Graph two or three values of the velocity vector function. What are the maximum velocity, altitude, and range of the projectile?

3. Using Example 5.2.3, find the gravitational intensity at a point 100,000 miles from the earth's center. Find the magnitude of the gravitational intensity at that point.

4. Using Example 5.2.3, find the gravitational intensity at a point 25,000 miles from the earth's center. Find the magnitude of the gravitational intensity at that point.

5. Using Example 5.2.4, evaluate and graph $\mathbf{v} = \overrightarrow{F(P)}$ at the points $P_1(\frac{1}{4}, 4)$, $P_2(1, 1)$, $P_3(4, \frac{1}{4})$. Find the magnitude of the velocity of the fluid at each of these points.

6. Using Example 5.2.4, evaluate and graph $\mathbf{v} = \overrightarrow{F(P)}$ at the points $P_1(\frac{1}{2}, 8)$, $P_2(2, 2)$, $P_3(8, \frac{1}{2})$. Find the magnitude of the velocity of the fluid at each of these points.

7. A thin rectangular plate 8 cm wide with a semi-infinite length has both surfaces insulated. If the temperature is kept at 0° C along the two long edges $x = 0$ and $x = 8$; at $100 \sin \pi x/8$ for $0 < x < 8$ along the bottom edge $y = 0$; and if the temperature at a point tends to zero as the distance between the point and the edge $y = 0$ tends to infinity, then the steady-state temperature at any point P in the plate is given by the following scalar point function:

$$u = \left\{ (T, P)/T = \left(100 \sin \frac{\pi x}{8} \right) \cdot e^{-\frac{\pi y}{8}}, \text{ where } P(x, y) \text{ is a point such that,} \right.$$

$$\left. 0 < x < 8, y > 0. \right\}$$

Locate the plate in a coordinate system as indicated above and find the temperature $T = u(P)$ for $P_1(4, 8/\pi)$,* $P_2(\frac{4}{3}, 16/\pi)$, and $P_3(2, 8)$.

* Use a table of exponentials to evaluate $e^{-\frac{\pi y}{8}}$.

references

1. Brand, L., *Vector Analysis*, John Wiley, 1957.
2. Davis, H., *Introduction to Vector Analysis*, Allyn and Bacon, 1961.
3. Halliday and Resnick, *Physics*, John Wiley, 1961.
4. Hay, G. E., *Vector and Tensor Analysis*, Dover, 1953.
5. Kaplan, W., *Advanced Calculus*, Addison Wesley, 1953.
6. Nara, H. R. (Editor), *Vector Mechanics for Engineers*, Parts I and II, John Wiley, 1962.
7. Lass, Harry, *Vector and Tensor Analysis*, McGraw-Hill, 1950.
8. Phillips, H. B., *Vector Analysis*, John Wiley, 1933.
9. Richards, Sears et al. *Modern University Physics*, Addison Wesley, 1960.
10. Schuster, S., *Elementary Vector Geometry*, John Wiley, 1962.
11. Schwartz, Green, and Rutledge, *Vector Analysis with Applications*, Harper, 1960.
12. Sears, *Mechanics, Wave Motion, and Heat*, Addison Wesley, 1958.
13. Taylor, A. E., *Advanced Calculus*, Ginn, 1955.
14. Weatherburn, C. E., *Elementary Vector Analysis*, George Bell and Sons, 1921, 1928
15. Wills, A. P., *Vector Analysis with an Introduction to Tensor Analysis*, Dover, 1958.
16. Wilson and Gibbs, *Vector Analysis*, Yale University Press, 1947.

answers to problem sets

PROBLEM SET #1: A. General

1.

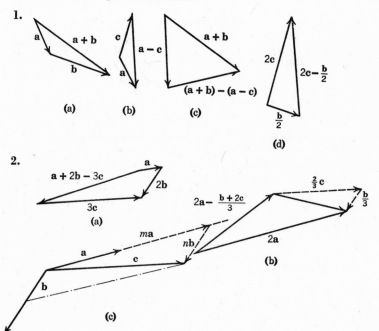

(a)

(b)

(c)

(d)

2.

(a)

(b)

(c)

3.

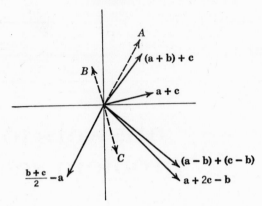

4.

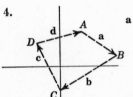

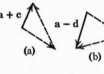

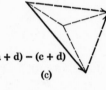

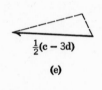

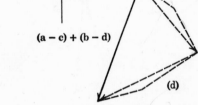

5. A vector with magnitude one (i.e., a unit vector) having the same direction as **n**.

6. $|\overrightarrow{AB}| = 5$, direction: 53°8′.   Magnitude of $\dfrac{\overrightarrow{AB}}{|\overrightarrow{AB}|}$ is 1.

7. $\overrightarrow{BC} = \mathbf{c} - \mathbf{b},\ \overrightarrow{BD} = \mathbf{d} - \mathbf{b},\ \overrightarrow{CD} = \mathbf{d} - \mathbf{c}.$

8. $\overrightarrow{AB} = \overrightarrow{OB} - \overrightarrow{OA},\ \overrightarrow{AC} = \overrightarrow{OC} - \overrightarrow{OA},\ \overrightarrow{AD} = \overrightarrow{OD} - \overrightarrow{OA}.$

9. $\overrightarrow{AC} = \mathbf{a} + \mathbf{b},\ \ \overrightarrow{AM} = \frac{1}{2}(\mathbf{a} + \mathbf{b}),\ \ \overrightarrow{BD} = \mathbf{b} - \mathbf{a},\ \ \overrightarrow{BM} = \frac{1}{2}(\mathbf{b} - \mathbf{a}),$
   $\overrightarrow{AN} = \frac{1}{2}\mathbf{a},\ \overrightarrow{DN} = \frac{1}{2}\mathbf{a} - \mathbf{b},\ \overrightarrow{MN} = -\frac{1}{2}\mathbf{b}.$

10. The sum vector is the main diagonal of a cube with 2 unit edges. Its magnitude is $2\sqrt{3}$.

11. Left to right: $(x - z)\mathbf{a} + (y - w)\mathbf{b} = \mathbf{0}$ and Theorem 1.3.3.
    Right to left: $x - z = 0$ and $y - w = 0,\ (x - z)\mathbf{a} + (y - w)\mathbf{b} = \mathbf{0}.$

12.

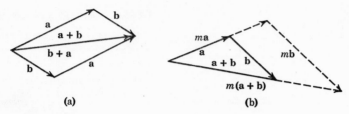

(a)   (b)

13. The shortest distance between two points is a straight line.

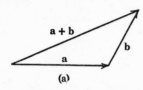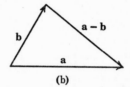

(a)   (b)

14. Theorem 1.3.4: Let $\mathbf{v}_1 \| \mathbf{v}_2$; then $\mathbf{v}_1 = k\mathbf{v}_2$ and $\mathbf{v}_1$, $\mathbf{v}_2$ are not linearly independent.   Theorem 1.3.6: Similar proof using contrapositive form.

B. Geometric

1. Start with any of the other vertices or use a different point Q.

2. $\overrightarrow{PM_1} = \frac{1}{2}(\overrightarrow{PA} + \overrightarrow{PB})$, etc.   See Example 1.2.1.

3. $\overrightarrow{OP} = \mathbf{b} + \dfrac{\alpha}{\alpha + \beta}(\mathbf{a} - \mathbf{b})$, etc.

4.

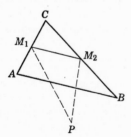

Show that $\overrightarrow{M_1M_2} = \frac{1}{2}\overrightarrow{AB}$

$$\overrightarrow{M_1M_2} = \overrightarrow{PM_2} - \overrightarrow{PM_1}$$
$$= \frac{1}{2}(\overrightarrow{PB} - \overrightarrow{PA}) = \frac{1}{2}\overrightarrow{AB}.$$

5.

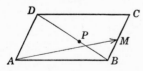

Let P be such that $\dfrac{BP}{BD} = \frac{1}{3}$ and M be the midpoint of BC.

$$\overrightarrow{AP} = \frac{1}{3}\overrightarrow{AD} + \frac{2}{3}\overrightarrow{AB} = \frac{2}{3}\overrightarrow{BM}$$
$$+ \frac{2}{3}\overrightarrow{AB} = \frac{2}{3}\overrightarrow{AM}.$$

6.

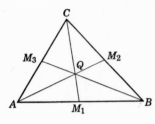

(i) Choose point P and consider $\overrightarrow{PM_1}$, $\overrightarrow{PM_2}$, $\overrightarrow{PM_3}$ (see Example 1.2.1).

(ii) Let Q_1, Q_2, Q_3 divide the medians as required.

(iii) Use the results of Problem B-3 to express $\overrightarrow{PQ_1}$, $\overrightarrow{PQ_2}$, $\overrightarrow{PQ_3}$.

(iv) Observe that $\overrightarrow{PQ_1} = \overrightarrow{PQ_2} = \overrightarrow{PQ_3}$.

7.

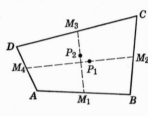

Let P_1 and P_2 be the midpoints and choose a point O. Show that $\overrightarrow{OP_1} = \overrightarrow{OP_2}$.

(*Hint:* $\overrightarrow{OP_1} = \frac{1}{2}\overrightarrow{OM_4} + \frac{1}{2}\overrightarrow{OM_2}$
$= \frac{1}{2}[\frac{1}{2}\overrightarrow{OA} + \frac{1}{2}\overrightarrow{OD}]$
$\qquad + \frac{1}{2}[\frac{1}{2}\overrightarrow{OC} + \frac{1}{2}\overrightarrow{OB}]$

8. Proof is similar to B-7. Parallelogram.

9. (a) 1-space: 2 points, each 2 units from O.

         2-space: A circle with center O and radius 2 units.

         3-space: A sphere with center O and radius 2 units.

   (b) 1-space: A four-unit line segment with O as its midpoint and excluding the endpoints.

         2-space: The interior of the circle with center O and radius 2 units.

         3-space: The interior of the phere with center O and radius 2 units.

   (c) 1-space: as (b) but also including the endpoints.

         2-space: as (b) but also including the circle.

         3-space: as (b) but also including the sphere.

   (d) 1-space: the line excluding (c).

         2-space: The exterior of the circle with center O and radius 2 units.

         3-space: The exterior of the sphere with center O and radius 2 units.

10. 1-space:

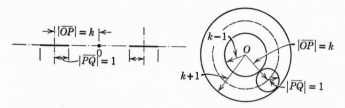

2-space: The ring between the concentric circles.

3-space: The locus is a spherical shell, 2 units thick, with center at O, inner radius $k - 1$, and outer radius $k + 1$.

11. (a) The perpendicular bisector of the line segment AB.

(b) An ellipse with foci at A and B.

(c) No points as locus—the difference of two sides of a triangle is always less than the third side.

12. Q is a point such that the quadrilateral $ABQC$ is a parallelogram.

13. $\overrightarrow{AP} = k(\overrightarrow{OB} - \overrightarrow{OA})$ where k takes all real values.

14. $\overrightarrow{P_0P} = m\left[\dfrac{|\mathbf{b}|}{|\mathbf{a}|}\,\mathbf{a} \pm \mathbf{b}\right]$ where m assumes all real values.

15. $\overrightarrow{OP} = \alpha\overrightarrow{OA} + \beta\overrightarrow{OB}$ where α and β assume all real values.

16. (a) $\overrightarrow{AP} = t\overrightarrow{AB}$ where t takes all real values.

(b) $|\overrightarrow{AP}|^2 + |\overrightarrow{AB}|^2 = |\overrightarrow{PA}|^2$

(c) $|t\overrightarrow{AB}|^2 + 1 = |\overrightarrow{AP}|^2$ where $0 \le t \le |\overrightarrow{AB}|$.

17. $|\overrightarrow{AP}|^2 + |\overrightarrow{BP}|^2 = |\overrightarrow{AB}|^2$.

18. $|\overrightarrow{AP}|^2 + |\overrightarrow{BP}|^2 = |\overrightarrow{AB}|^2$.

C. Physics

1. (a) Vector   (b) scalar   (c) scalar   (d) vector   (e) scalar
(f) vector

2. (a) Scalar   (b) vector   (c) scalar   (d) vector   (e) scalar
(f) vector

3. Direction: 14.6° North of East.   Magnitude: 14 miles.

4. Direction: 65.5° East of North.   Magnitude: 6.22 miles.

5. Approximately 1732 miles North and 1000 miles East.

6. $(1950, -1950\sqrt{3})$

7. 36°53′30″ North of West, 1297.3 miles from its starting point.

8. The velocity is approximately 313 mph in the direction 22°54′ West of North.

9. Compass heading: 18°12′36″ North of the East-West line. The speed with respect to the land will be 7.6 knots.
10. Angle: 21°48′ Length: 11.55 miles.
11.

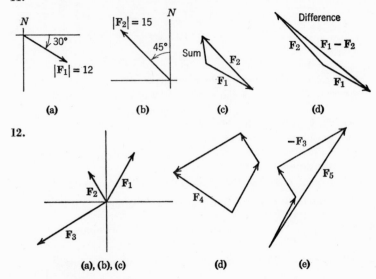

(a)   (b)   (c)   (d)

12.

(a), (b), (c)   (d)   (e)

13. A force of 5 gm at 270°.
14. A force of 59.5 dynes at 78°45′.
15. 38 lb at 0° and 75 $\sqrt{3}$ lb at 90°.
16. 100 lb at 90° and 200 $\sqrt{3}$ lb at 180°.

PROBLEM SET #2.1: A. General

1. Prove by considering eight cases. See Figure 2.1.3 for four cases.
2. Use the definition: $\mathbf{a} \cdot \mathbf{b} = |\mathbf{a}|\,|\mathbf{b}|\cos\theta$; $\mathbf{a}$ and $\mathbf{b}$ are nonzero vectors.
3. $\mathbf{a} \cdot \mathbf{b} = |\mathbf{a}|\,|\mathbf{b}|\cos\theta = |\mathbf{b}|\,|\mathbf{a}|\cos\theta = \mathbf{b} \cdot \mathbf{a}$.
4. $\mathbf{a} \cdot (\mathbf{b} + \mathbf{c}) = |\mathbf{a}|\,\mathrm{comp}_a\,(\mathbf{b} + \mathbf{c}) = |\mathbf{a}|\,[\mathrm{comp}_a\,\mathbf{b} + \mathrm{comp}_a\,\mathbf{c}]$
   $= |\mathbf{a}|\,\mathrm{comp}_a\,\mathbf{b} + |\mathbf{a}|\,\mathrm{comp}_a\,\mathbf{c} = \mathbf{a} \cdot \mathbf{b} + \mathbf{a} \cdot \mathbf{c}$.
5. Show that all three expressions equal $h|\mathbf{a}|\,|\mathbf{b}|\cos\measuredangle(\mathbf{a}, \mathbf{b})$.
6. $\mathbf{a} \cdot \mathbf{a} = |\mathbf{a}|\,|\mathbf{a}|\cos\measuredangle(\mathbf{a}, \mathbf{a}) = |\mathbf{a}|\,|\mathbf{a}|\,1 = |\mathbf{a}|^2$.
7. $(\mathbf{a} + \mathbf{b}) \cdot (\mathbf{c} + \mathbf{d}) = (\mathbf{a} + \mathbf{b}) \cdot \mathbf{c} + (\mathbf{a} + \mathbf{b}) \cdot \mathbf{d}$ Alg. Lw. 2.1.2
   $= \mathbf{c} \cdot (\mathbf{a} + \mathbf{b}) + \mathbf{d} \cdot (\mathbf{a} + \mathbf{b})$ Alg. Lw. 2.1.1
   $= \mathbf{c} \cdot \mathbf{a} + \mathbf{c} \cdot \mathbf{b} + \mathbf{d} \cdot \mathbf{a} + \mathbf{d} \cdot \mathbf{b}$ Alg. Lw. 2.1.2
   $= \mathbf{c} \cdot \mathbf{a} + \mathbf{d} \cdot \mathbf{a} + \mathbf{c} \cdot \mathbf{b} + \mathbf{d} \cdot \mathbf{b}$ Alg. Lw. 1.2.1
   $= \mathbf{a} \cdot \mathbf{c} + \mathbf{a} \cdot \mathbf{d} + \mathbf{b} \cdot \mathbf{c} + \mathbf{b} \cdot \mathbf{d}$ Alg. Lw. 2.1.1

8. $|\mathbf{a}| = \sqrt{\mathbf{a} \cdot \mathbf{a}}$.

9. $(\mathbf{v} \cdot \mathbf{i})\mathbf{i} + (\mathbf{v} \cdot \mathbf{j})\mathbf{j} + (\mathbf{v} \cdot \mathbf{k})\mathbf{k} = (\text{comp}_i \mathbf{v})\mathbf{i} + (\text{comp}_j \mathbf{v})\mathbf{j} + (\text{comp}_k \mathbf{v})\mathbf{k} = \mathbf{v}$.

10. $(\mathbf{a} \cdot \mathbf{b})(\mathbf{a} \cdot \mathbf{b}) = |\mathbf{a}|^2|\mathbf{b}|^2 \cos^2 \measuredangle(\mathbf{a}, \mathbf{b}) \le |\mathbf{a}|^2|\mathbf{b}|^2$ where the equality holds for $\measuredangle(\mathbf{a}, \mathbf{b}) = 0$.

B. Geometric

1.

Given: $\measuredangle A = \measuredangle B$, M is midpoint of AB.
Show that $CM \perp AB$
$\overrightarrow{CM} = \tfrac{1}{2}\overrightarrow{CB} + \tfrac{1}{2}\overrightarrow{CA}$, $\overrightarrow{AB} = \overrightarrow{CB} - \overrightarrow{CA}$
Complete by showing that $\overrightarrow{CM} \cdot \overrightarrow{AB} = 0$.

2.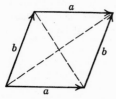

Given: $|\mathbf{a}| = |\mathbf{b}|$. Show that
$(\mathbf{a} - \mathbf{b}) \cdot (\mathbf{b} - \mathbf{a}) = 0$

3.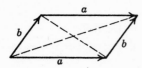

Show that $(\mathbf{a} - \mathbf{b})^2 + (\mathbf{b} - \mathbf{a})^2 = 2\mathbf{a}^2 + 2\mathbf{b}^2$

4.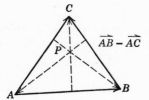

Given: $\overrightarrow{PC} \perp \overrightarrow{AB}$, $\overrightarrow{PB} \perp \overrightarrow{AC}$
Show: $\overrightarrow{PA} \perp \overrightarrow{BC}$
Sketch: $\overrightarrow{AB} \cdot \overrightarrow{PC} = \overrightarrow{AC} \cdot \overrightarrow{PB} = 0$
$\overrightarrow{PB} + (\overrightarrow{AB} - \overrightarrow{AC}) = \overrightarrow{PC}$
$\qquad\qquad\qquad = \overrightarrow{PA} + \overrightarrow{AC}$
$\overrightarrow{PB} = \overrightarrow{PA} + \overrightarrow{AB}$
then $(\overrightarrow{AB} - \overrightarrow{AC}) \cdot \overrightarrow{PA} = 0$.

5.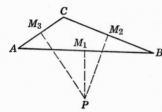

Let perpendicular bisectors of AC and BC intersect at P. Show that
$$\overrightarrow{PM_1} \perp \overrightarrow{AB},$$
$$(\tfrac{1}{2}\overrightarrow{PA} + \tfrac{1}{2}\overrightarrow{PB}) \cdot (\overrightarrow{PB} - \overrightarrow{PA}) = 0.$$

6.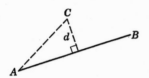

$$d = \sqrt{|\overrightarrow{AC}|^2 - \frac{\overrightarrow{AC} \cdot \overrightarrow{AB}}{|\overrightarrow{AB}|}}$$

7.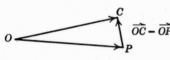

$\overrightarrow{OC} - \overrightarrow{OP}$

The locus is a plane perpendicular to $\overrightarrow{OC}$.

8.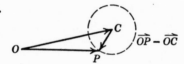

$\overrightarrow{OP} - \overrightarrow{OC}$

The locus is a sphere with radius r and center C.

C. Physics
1. 20 ft lb.
2. 20 ft lb.
3. Zero.
4. (a) 120 ft lb   (b) −18 ft lb   (c) 102 ft lb
  (d) (a) $10 \sqrt{3}$ ft lb   (b) −18 ft lb   (c) $(10 \sqrt{3} - 18)$ ft lb
5. 225 ft lb
6. $(5 \sqrt{3} - 4)50$ ft lb
7. $\overrightarrow{OX}$ is the vector from $(0, 0)$ to $(4, 0)$.
  $\overrightarrow{OY}$ is the vector from $(0, 0)$ to $(0, 2)$.

8.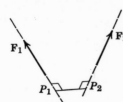

$\mathbf{F} = \mathbf{F}_1 + \mathbf{F}_2,\ \mathbf{F}_1 \cdot \overrightarrow{P_1P_2} = 0,$
$\mathbf{F}_2 \cdot \overrightarrow{P_1P_2} = 0.$
Show $(\mathbf{F}_1 + \mathbf{F}_2) \cdot \overrightarrow{P_1P_2} = 0.$

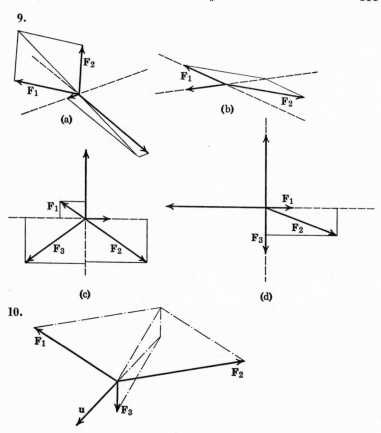

9.

(a)

(b)

(c)

(d)

10.

PROBLEM SET #2.2: A. General

1. $\mathbf{a} + \mathbf{b} = (a_x\mathbf{i} + a_y\mathbf{j} + a_z\mathbf{k}) + (b_x\mathbf{i} + b_y\mathbf{j} + b_z\mathbf{k})$
$= (a_x\mathbf{i} + b_x\mathbf{i}) + (a_y\mathbf{j} + b_y\mathbf{j}) + (a_z\mathbf{k} + b_z\mathbf{k})$
$= (a_x + b_x)\mathbf{i} + (a_y + b_y)\mathbf{j} + (a_z + b_z)\mathbf{k}.$

2. $h\mathbf{a} = h(a_x\mathbf{i} + a_y\mathbf{j} + a_z\mathbf{k}) = h(a_x\mathbf{i}) + h(a_y\mathbf{j}) + h(a_z\mathbf{k})$
$= (ha_x)\mathbf{i} + (ha_y)\mathbf{j} + (ha_z)\mathbf{k}.$

3. $\mathbf{a} \cdot \mathbf{b} = (a_x\mathbf{i} + a_y\mathbf{j} + a_z\mathbf{k}) \cdot (b_x\mathbf{i} + b_y\mathbf{j} + b_z\mathbf{k})$
$= (a_x\mathbf{i} + a_y\mathbf{j} + a_z\mathbf{k}) \cdot b_x\mathbf{i} + (a_x\mathbf{i} + a_y\mathbf{j} + a_z\mathbf{k}) \cdot b_y\mathbf{j} + \cdots$
$= (a_x\mathbf{i}) \cdot b_x\mathbf{i} + (a_y\mathbf{j}) \cdot b_x\mathbf{i} + (a_z\mathbf{k}) \cdot b_x\mathbf{i} + (a_x\mathbf{i}) \cdot b_y\mathbf{j} + \cdots$
$= (a_xb_x)(\mathbf{i} \cdot \mathbf{i}) + (a_yb_x)(\mathbf{j} \cdot \mathbf{i}) + (a_zb_x)(\mathbf{k} \cdot \mathbf{i}) + \cdots$
$= a_xb_x + a_yb_y + a_zb_z$ since $\mathbf{i} \cdot \mathbf{i} = \mathbf{j} \cdot \mathbf{j} = \mathbf{k} \cdot \mathbf{k} = 1$ and
$\mathbf{i} \cdot \mathbf{j} = \mathbf{j} \cdot \mathbf{k} = \mathbf{k} \cdot \mathbf{i} = 0.$

4. See Figure 2.3.3: $\overrightarrow{P_1P_2} = \overrightarrow{OP_2} - \overrightarrow{OP_1}$
$$= (x_2\mathbf{i} + y_2\mathbf{j} + z_2\mathbf{k}) - (x_1\mathbf{i} + y_1\mathbf{j} + z_1\mathbf{k})$$
$$= (x_2 - x_1)\mathbf{i} + (y_2 - y_1)\mathbf{j} + (z_2 - z_1)\mathbf{k}.$$

$$|\overrightarrow{P_1P_2}| = \sqrt{\overrightarrow{P_1P_2}^2} = \sqrt{(x_2 - x_1)^2 + (y_2 - y_1)^2 + (z_2 - z_1)^2}$$

$$|\overrightarrow{P_1P_2}| = \sqrt{(1 + 3)^2 + (-2 + 1)^2 + (3 - 1)^2} = \sqrt{21}.$$

5. $\cos \angle(\mathbf{a}, \mathbf{b}) = \dfrac{\mathbf{a} \cdot \mathbf{b}}{|\mathbf{a}|\,|\mathbf{b}|} = \dfrac{a_xb_x + a_yb_y + a_zb_z}{\sqrt{a_x^2 + a_y^2 + a_z^2}\,\sqrt{b_x^2 + b_y^2 + b_z^2}}$

by the definition $\mathbf{a} \cdot \mathbf{b} = |\mathbf{a}|\,|\mathbf{b}|\cos \angle(\mathbf{a}, \mathbf{b})$ and Theorems 2.3.8, 2.3.9.

6. $\left|\dfrac{\mathbf{a}}{|\mathbf{a}|}\right| = \sqrt{\left(\dfrac{a_x}{|\mathbf{a}|}\right)^2 + \left(\dfrac{a_y}{|\mathbf{a}|}\right)^2 + \left(\dfrac{a_z}{|\mathbf{a}|}\right)^2} = \dfrac{1}{|\mathbf{a}|}\sqrt{a_x^2 + a_y^2 + a_z^2}$

$$= \dfrac{|\mathbf{a}|}{|\mathbf{a}|} = 1.$$

$\dfrac{\mathbf{a}}{|\mathbf{a}|}$ has the same sense as $\mathbf{a}$, since $\dfrac{1}{|\mathbf{a}|}$ is a positive scalar.

7. See Theorems 2.4.1 and 2.3.12.

8. Definition 2.4.2 and $\mathbf{a}\|\mathbf{b} \Leftrightarrow \mathbf{a} = k\mathbf{b}$.

9. $\mathbf{a} = k\dfrac{\mathbf{a}}{|\mathbf{a}|}$.

10. $\operatorname{comp}_a \mathbf{c} = |\mathbf{c}|\cos \angle(\mathbf{a}, \mathbf{c}) = \dfrac{|\mathbf{a}|\,|\mathbf{c}|}{|\mathbf{a}|}\cos \angle(\mathbf{a}, \mathbf{c}) = \dfrac{\mathbf{a} \cdot \mathbf{c}}{|\mathbf{a}|}$.

11. (a) $\sqrt{14}$ (b) $6\mathbf{i} + 4\mathbf{j} + \mathbf{k}$ (c) $4\mathbf{i} + \mathbf{j} + 5\mathbf{k}$ (d) 4

(e) $-\tfrac{13}{6}\sqrt{3}$ (f) 3 (g) $4\mathbf{i} + 6\mathbf{j} - 2\mathbf{k}$ (h) $-\dfrac{13}{2\sqrt{42}}$

(i) $\dfrac{3}{\sqrt{14}}\mathbf{i} + \dfrac{2}{\sqrt{14}}\mathbf{j} - \dfrac{1}{\sqrt{14}}\mathbf{k}$

(j) Direction numbers: 3, 2, −1

Direction cosines: $\dfrac{3}{\sqrt{14}}, \dfrac{2}{\sqrt{14}} - \dfrac{1}{\sqrt{14}}$.

12. (a) $\sqrt{6}$ (b) $-2\mathbf{i} + \mathbf{j} + 3\mathbf{k}$ (c) $-4\mathbf{i} - 2\mathbf{j} + 5\mathbf{k}$ (d) -5

(e) $\tfrac{5}{6}\sqrt{6}$ (f) 1 (g) $-4\mathbf{i} + 6\mathbf{k}$ (h) $\tfrac{5}{42}\sqrt{21}$

(i) $\dfrac{1}{\sqrt{6}}\mathbf{i} + \dfrac{2}{\sqrt{6}}\mathbf{j} - \dfrac{1}{\sqrt{6}}\mathbf{k}$.

(j) Direction numbers: 1, 2, −1

  Direction cosines: $\dfrac{1}{\sqrt{6}}, \dfrac{2}{\sqrt{6}}, -\dfrac{1}{\sqrt{6}}$.

13. (a) $-\mathbf{i} - 3\mathbf{k}$ (b) $\sqrt{10}$ (c) 7 (d) $\dfrac{7}{\sqrt{95}}$ (e) $\dfrac{2}{\sqrt{5}}, \dfrac{1}{\sqrt{5}}, 0$.

14. (a) $-2\mathbf{i} + \mathbf{j} + \mathbf{k}$ (b) $\sqrt{6}$ (c) 5 (d) $\dfrac{5}{2\sqrt{7}}$ (e) $\dfrac{1}{\sqrt{2}}, \dfrac{-1}{\sqrt{2}}, 0$.

15. $3\mathbf{i} + 4\mathbf{k}$.

16. $\dfrac{1}{\sqrt{42}}, \dfrac{5}{\sqrt{42}}, \dfrac{-4}{\sqrt{42}}$; two sets.

17. $\alpha = \tfrac{7}{8}, \beta = -\tfrac{5}{8}$; not collinear or proportional, no.

18. $\alpha = 1, \beta = -2, \gamma = -3$; **a**, **b**, and **c** are not coplanar.

B. Geometric

1. (a) $P\left(0, -\dfrac{\gamma}{\beta}\right)$ and $Q\left(-\dfrac{\gamma}{\alpha}, 0\right)$ are on $\alpha x + \beta y + \gamma = 0$. Form

  $\overrightarrow{QP} = \dfrac{\gamma}{\alpha}\mathbf{i} - \dfrac{\gamma}{\beta}\mathbf{j}$; then $\mathbf{N} \cdot \overrightarrow{QP} = 0$. Hence **N** is perpendicular

  to the line.

  (b) $\mathbf{U} = \tfrac{2}{13}\mathbf{i} - \tfrac{3}{13}\mathbf{j}$.

  (c) $d = |\overrightarrow{QP}| \cos \not\prec (\overrightarrow{QP}, \mathbf{N}) = |\overrightarrow{QP}| \dfrac{\overrightarrow{QP} \cdot \mathbf{N}}{|\overrightarrow{QP}|\,|\mathbf{N}|}$.

  (d) $\mathbf{N} = a\mathbf{i} + b\mathbf{j}$ is the normal to $ax + by + c = 0$.

  $\overrightarrow{QP} = (x_1 - x_0)\mathbf{i} + (y_1 - y_0)\mathbf{j}$ for $Q(x_0, y_0)$ on the line.

  Then from (c) we obtain the conclusion by substitution.

  (e) $d = \tfrac{2}{3}\sqrt{3}$.

2. $\sqrt{5}$.

3. (a) (i) $\overrightarrow{P_0Q} \cdot \mathbf{N} = 0$ where Q is an arbitrary point on the plane.

  (ii) $A(x - x_0) + B(y - y_0) + C(z - z_0) = 0$

   or $Ax + By + Cz + D = 0$

   where $D = -(Ax_0 + By_0 + Cz_0)$.

  (b) $-(x - 2) + 2(y - 1) + 2(z - 5) = 0$

   or $x - 2y - 2z + 10 = 0$.

4. $2(x - 1) + (y + 1) - (z - 3) = 0$ or $2x + y - z + 2 = 0$.

5. $\dfrac{3}{\sqrt{38}}\mathbf{i} - \dfrac{2}{\sqrt{38}}\mathbf{j} + \dfrac{5}{\sqrt{38}}\mathbf{k}$.

6. $\theta = \cos^{-1} \dfrac{10}{\sqrt{847}} = 69°45'$ approximately.

7. (a) $d = \left| \dfrac{\mathbf{N} \cdot \overrightarrow{QP}}{|\mathbf{N}|} \right|$ where Q is on the plane and $\mathbf{N} = A\mathbf{i} + B\mathbf{j} + C\mathbf{k}$.

  (b) $d = \left| \dfrac{Ax_1 + By_1 + Cz_1 + D}{\sqrt{A^2 + B^2 + C^2}} \right|.$

8. (a) $\sqrt{14}$ (b) $\dfrac{25}{\sqrt{14}}.$

9. (a) (i) $|\overrightarrow{P_0P}| = r$ or $\overrightarrow{P_0P}^2 = r^2$

  (ii) $(x - x_0)^2 + (y - y_0)^2 + (z - z_0)^2 = r^2$

  (b) $(x + 2)^2 + (y + 1)^2 + (z - 5)^2 = 64.$

10. $3x - 2y + 5z + \sqrt{38} = 0$ or $3x - 2y + 5z - \sqrt{38} = 0.$

11. $\mathbf{b} \cdot \mathbf{c} = 0$ and $\mathbf{a} = \mathbf{c} - \mathbf{b}.$

12. $\cos(\alpha_1 - \alpha_2) = \cos \measuredangle (\mathbf{a}, \mathbf{b}) = \dfrac{\mathbf{a} \cdot \mathbf{b}}{|\mathbf{a}| \, |\mathbf{b}|}$

$= \cos \alpha_1 \cos \alpha_2 + \cos \beta_1 \cos \beta_2$

$= \cos \alpha_1 \cos \alpha_2 + \sin \alpha_1 \sin \alpha_2.$

$\cos(\alpha_1 + \alpha_2)$ follows by substituting $-\alpha_2$ for $\alpha_2.$

13. The locus is a circle formed by the intersection of a cone (60° with respect to the y-axis) and a sphere of radius 5 with center at the origin.

14. $\theta = \cos^{-1} \dfrac{1}{\sqrt{3}} = 54°44'$ approximately.

15. Find a vector $\mathbf{u} = \alpha\mathbf{i} + \beta\mathbf{j} + \gamma\mathbf{k}$ such that $\mathbf{u} \cdot \mathbf{a} = 0$ and $\mathbf{u} \cdot \mathbf{b} = 0$. Is $\mathbf{u}$ unique?

16. $\mathbf{a} - \mathbf{b} = 3\mathbf{i} - 2\mathbf{j}; \mathbf{k} \cdot (\mathbf{a} - \mathbf{b}) = 0$; therefore $\mathbf{a} - \mathbf{b}$ is perpendicular to the z-axis. $|\mathbf{a} - \mathbf{b}| = \sqrt{13}.$

C. Physics

  1. Work done $= -8$ units.

  2. Work done $= -8$ units.

  3. The work done by the constant force $\mathbf{F}$ in displacing a particle from one point to another is independent of the path joining the two points.

  4. Consider the projection of a unit area perpendicular to the flow lines. Then $\mathbf{V} \cdot \mathbf{N} = |\mathbf{V}| \, |\mathbf{N}| \cos \measuredangle (\mathbf{V}, \mathbf{N})$

$= (|\mathbf{V}| \cos \theta)$ (unit area)

$=$ volume per unit time.

PROBLEM SET #3.1: A. General

1. (a) $-i - 3j - 2k$ (b) $-2j - k$ (c) $i + 3j + 2k$
   (d) $-i - 5j - 3k$ (e) $-10i + 4j - 3k$
2. (a) $i - 2j + 4k$ (b) $i + j + k$.   (c) $-i + 2j - 4k$
   (d) $2i - j + 5k$ (e) $3i + 6j$.
3. $-2i + 9j + 3k$.
4. $3i - 2j + k$.

5. (a) $\dfrac{1}{\sqrt{59}} (3i + j + 7k)$

   (b) $\sin \theta = \dfrac{|a \times b|}{|a| |b|} = \dfrac{\sqrt{59}}{\sqrt{6} \sqrt{14}} = \sqrt{\frac{59}{84}}$.

6. (a) $\dfrac{1}{\sqrt{61}} (6i - 4j + 3k)$ (b) $\sin \theta = \sqrt{\frac{61}{65}}$.

7. $(a + b) \times (a - b) = (a + b) \times a - (a + b) \times b$
   $$= -a \times (a + b) + b \times (a + b)$$
   $$= -a \times b + b \times a = -a \times b - a \times b$$
   $$= -2(a \times b) = -2a \times b.$$

8. $a \times (a + b + c) = 0$, $a \times b + a \times c = 0$, $a \times b - c \times a = 0$, $a \times b = c \times a$. Now show that $a \times b = b \times c$ follows from $b \times (a + b + c) = 0$.

9. Let $a \times b = a \times c$, $a \neq 0$. $a \times b - a \times c = 0$, $a \times (b - c) = 0$. Then a and $(b - c)$ are collinear or $b - c = 0$. Clearly $b - c$ can be a vector collinear with a without $b = c$.

10. $b = c + \alpha a$ given so that $a \times b = a \times (c + \alpha a) = a \times c + a \times \alpha a = a \times c$.

11. (i) $a \cdot b = a \cdot c$ implies $a \cdot (b - c) = 0$, $a \perp (b - c)$.
    (ii) $a \times b = a \times c$ implies $a \times (b - c) = 0$, $a \| (b - c)$.
    (iii) $a \neq 0$
    (iv) Thus $b - c = 0$ and $b = c$.

12. $a \times (b \times c)$ is a vector in the plane parallel to b and c. If a, b, and c are noncoplanar, then the triple products could not be the same vector.
    (*Note:* A counter example would do.)

13. Expand both sides in component form and note that the corresponding components are equal.

14. Follows directly from problem 13 by letting $a = c$ and $b = d$.

B. Geometric

1. $[a \times b = 0] \Leftrightarrow \dfrac{a_x}{b_x} = \dfrac{a_y}{b_y} = \dfrac{a_z}{b_z} \Leftrightarrow a \| b \Leftrightarrow [a$ and b collinear].
2. b and c since $b \times c = 0$.

3. (a) By problem 1 the vectors $\overrightarrow{P_1P_2}$ and $\overrightarrow{P_1P}$ are collinear. Since both vectors have a common initial point P_1, they must lie on the same line. Hence P lies on the line through P_1 and P_2. The reasoning is reversible.

(b) Write:

$$\overrightarrow{P_1P_2} \times \overrightarrow{P_1P} = \begin{vmatrix} \mathbf{i} & \mathbf{j} & \mathbf{k} \\ x_2 - x_1 & y_2 - y_1 & z_2 - z_1 \\ x - x_1 & y - y_1 & z - z_1 \end{vmatrix} = \mathbf{0}.$$

Expand and set components equal to zero.

(c) Set $y = y_1$, $\dfrac{x - x_1}{x_2 - x_1} = \dfrac{z - z_1}{z_2 - z_1}$. A line parallel to the xz-plane.

(d) Set $y = y_1$, $z = z_1$, and x arbitrary (i.e., $x = kx_1$ for all real k). A line parallel to the x-axis.

4. (a) $\dfrac{x - 2}{-1} = \dfrac{y + 1}{-1} = \dfrac{z - 3}{-5}$

(b) $y = -7$, $\dfrac{x - 5}{-9} = \dfrac{z - 1}{2}$ (c) $x = -1$, $z = 3$.

5. (a) $\overrightarrow{P_1P}$ and $\mathbf{a}$ are collinear and thus their direction numbers are proportional and by definition a_x, a_y, a_z are also direction numbers of $\overrightarrow{P_1P}$.

(b) See 3(b) with appropriate substitutions.

(c) As in 3(c) and 3(d) above.

6. (a) $\dfrac{x - 2}{4} = \dfrac{y + 1}{-3} = \dfrac{z}{-3}$

(b) $z = 1$, $\dfrac{x - 3}{2} = \dfrac{y + 2}{-3}$ (c) $x = 3$, $y = -2$.

7. (a) $\overrightarrow{P_1P} = t\mathbf{a} \Leftrightarrow \overrightarrow{P_1P} \times \mathbf{a} = 0$

(b) $(x - x_1)\mathbf{i} + (y - y_1)\mathbf{j} + (z - z_1)\mathbf{k} = ta_x\mathbf{i} + ta_y\mathbf{j} + ta_z\mathbf{k}$;
$x - x_1 = ta_x$, $y - y_1 = ta_y$, $z - z_1 = ta_z$.

(c) If $x = x_1 + ta_x$, then $\dfrac{x - x_1}{a_x} = t$.

If $y = y_1 + ta_y$, then $\dfrac{y - y_1}{a_y} = t$.

If $z = z_1 + ta_z$, then $\dfrac{z - z_1}{a_z} = t$.

Thus $\dfrac{x - x_1}{a_x} = \dfrac{y - y_1}{a_y} = \dfrac{z - z_1}{a_z} = t$.

8. (a) $x = 2 - t$ (b) $x = 5 - 9t$ (c) $x = -1$

$\quad\quad$ $y = -1 - t$ $\quad\quad$ $y = -7$ $\quad\quad$ $y = 2 + 2t$

$\quad\quad$ $z = 3 - 5t$ $\quad\quad$ $z = 1 + 2t$ $\quad\quad$ $z = 3$

$\quad$ (d) $x = 2 - 4t$ (e) $x = 3 + 2t$ (f) $x = 3$

$\quad\quad$ $y = -1 + 3t$ $\quad\quad$ $y = -2 - 3t$ $\quad\quad$ $y = -2$

$\quad\quad$ $z = 3t$ $\quad\quad$ $z = 1$ $\quad\quad$ $z = 1 - 4t$

9. (a) First, if $(\overrightarrow{P_1P_2} \times \overrightarrow{P_1P_3}) \cdot \overrightarrow{P_1P} = 0$, we have $(\overrightarrow{P_1P_2} \times \overrightarrow{P_1P_3})$ is a vector perpendicular to $\overrightarrow{P_1P}$, so that P must lie in the plane determined by P_1, P_2, and P_3. If P lies in the plane determined by P_1, P_2, and P_3, then $\overrightarrow{P_1P}$ is perpendicular to $(\overrightarrow{P_1P_2} \times \overrightarrow{P_1P_3})$, so that $(\overrightarrow{P_1P_2} \times \overrightarrow{P_1P_3}) \cdot \overrightarrow{P_1P} = 0$.

$\quad$ (b) $\begin{vmatrix} \mathbf{i} & \mathbf{j} & \mathbf{k} \\ (x_2 - x_1) & (y_2 - y_1) & (z_2 - z_1) \\ (x_3 - x_1) & (y_3 - y_1) & (z_3 - z_1) \end{vmatrix} \cdot [(x - x_1)\mathbf{i} + (y - y_1)\mathbf{j} + (z - z_1)\mathbf{k}] = 0.$

$\quad$ Expand and rewrite in determinant form.

10. $18\mathbf{i} + 4\mathbf{j} + 6\mathbf{k}$.

11. $18x + 4y + 6z - 8 = 0$.

12. (a) $7\mathbf{i} - 8\mathbf{j} - 13\mathbf{k}$.

$\quad$ (b) $7(x - x_1) - 8(y - y_1) - 13(z - z_1) = 0$ where $P_1(x_1, y_1, z_1)$ is a fixed point in the plane.

13. (a) $7(x - 2) - 8(y - 1) - 13(z + 3) = 0$ and $7(x - 4) - 8y - 13(z - 4) = 0$.

$\quad$ (b) $d = \dfrac{69}{\sqrt{282}}$.

14. $\left(\dfrac{\mathbf{a} \times \mathbf{b}}{|\mathbf{a} \times \mathbf{b}|}\right) \cdot \overrightarrow{P_0P_1}$ gives the distance between the lines m_1 and m_2.

$\quad$ The distance between the two parallel planes is the same as the distance between the two lines.

15. The normal to the plane determined by P_1, P_2, and P_3 is the vector product. The unit normal **N** is thus

$$\mathbf{N} = \dfrac{\overrightarrow{P_1P_2} \times \overrightarrow{P_1P_3}}{|\overrightarrow{P_1P_2} \times \overrightarrow{P_1P_2}|} \text{ and } d = |\mathbf{N} \cdot \overrightarrow{P_1P_0}|.$$

16. $d = \dfrac{80}{\sqrt{376}}$.

17. See problem 14.

18. $d = \dfrac{2}{\sqrt{34}}$.

19. $d = \dfrac{81}{\sqrt{234}}.$

20. (a) Area $\triangle ABC = \frac{1}{2}\sqrt{51}$ (b) $\dfrac{17}{\sqrt{51}}$ (c) $d = \dfrac{19}{\sqrt{57}}.$

21. Area $= \frac{1}{2}\left|\overrightarrow{AB} \times \overrightarrow{AC}\right| = \frac{1}{2}\begin{Vmatrix} \mathbf{i} & \mathbf{j} & \mathbf{k} \\ x_2 - x_1 & y_2 - y_1 & z_2 - z_1 \\ x_3 - x_1 & y_3 - y_1 & z_3 - z_1 \end{Vmatrix}$

$= \frac{1}{2}\sqrt{[(y_2 - y_1)(z_3 - z_1) - (y_3 - y_1)(z_2 - z_1)]^2}$
$\overline{+ [(x_3 - x_1)(z_2 - z_1) - (x_2 - x_1)}$
$(z_3 - z_1)]^2 + [(x_2 - x_1)(y_3 - y_1) - (x_3 - x_1)(y_2 - y_1)]^2}.$

22. $\frac{1}{2}\sqrt{195}.$

23. Volume $= \mathbf{a} \cdot (\mathbf{b} \times \mathbf{c}) = \begin{vmatrix} a_x & a_y & a_z \\ b_x & b_y & b_z \\ c_x & c_y & c_z \end{vmatrix}.$

24. Note that the volume of a tetrahedron is equal to $\frac{1}{6}$ the volume of the parallelepiped from which it can be cut. $V = \frac{20}{6}$

25. Let $\mathbf{a}, \mathbf{b}, \mathbf{c}$ be the sides in vector form with $\mathbf{a} = \mathbf{c} - \mathbf{b}$.
$|\mathbf{b} \times \mathbf{c}| = |(\mathbf{c} - \mathbf{b}) \times (-\mathbf{b})| = |-(\mathbf{c} - \mathbf{b}) \times (-\mathbf{c})|.$
$bc \sin A = ab \sin C = ac \sin B$, and the conclusion follows.

26.

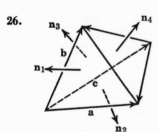

$\mathbf{n}_1 = \frac{1}{2}(\mathbf{a} \times \mathbf{b})$, $\mathbf{n}_2 = \frac{1}{2}(\mathbf{c} \times \mathbf{a})$, $\mathbf{n}_3 = \frac{1}{2}(\mathbf{b} \times \mathbf{c})$, $\mathbf{n}_4 = \frac{1}{2}[(\mathbf{b} - \mathbf{c}) \times (\mathbf{a} - \mathbf{c})]$. $\mathbf{n}_1 + \mathbf{n}_2 + \mathbf{n}_3 + \mathbf{n}_4 = \mathbf{0}$ by substitution and simplification.

C. Physics

1. $k_1\mathbf{N} \times \mathbf{a}_1 = k_2\mathbf{N} \times \mathbf{a}_2$, $(k_1|\mathbf{N}|\,|\mathbf{a}_1|\sin\alpha_1)\mathbf{E} = (k_2|\mathbf{N}|\,|\mathbf{a}_2|\sin\alpha_2)\mathbf{E}$ where $\mathbf{E}$ is a common unit vector, $k_1 \sin\alpha_1 = k_2 \sin\alpha_1$. (Reversible.)

2. Note that $\mathbf{v} = \mathbf{w} \times \mathbf{r}$ is perpendicular to both $\mathbf{w}$ and $\mathbf{r}$; thus $\mathbf{v}$ is tangent to the circle. $|\mathbf{v}| = |\mathbf{w} \times \mathbf{r}| = |\mathbf{w}|\,|\mathbf{r}|\sin\theta = w(\text{radius})$; thus $|\mathbf{v}| = $ velocity of a point on the circle.

3. (a) $|\overrightarrow{P_0P_1}|F \sin\theta = |\overrightarrow{P_0P_1} \times \mathbf{F}|.$ (b) $\sqrt{370}.$

4. $|\mathbf{r}|F \sin\theta = |\mathbf{r} \times \mathbf{F}|.$

PROBLEM SET #3.2: A. General

1. If **a**, **b**, **c** are coplanar, then $\mathbf{a} = k_1\mathbf{b} + k_2\mathbf{c}$ and $\mathbf{a} \cdot \mathbf{b} \times \mathbf{c}$ can be

expressed as $\begin{vmatrix} (k_1b_x + k_2c_x) & (k_1b_y + k_2c_y) & (k_1b_z + k_2c_z) \\ b_x & b_y & b_z \\ c_x & c_y & c_z \end{vmatrix} = 0.$

If $\mathbf{a} \cdot \mathbf{b} \times \mathbf{c} = 0$, then $\begin{vmatrix} a_x & a_y & a_z \\ b_x & b_y & b_z \\ c_x & c_y & c_z \end{vmatrix} = 0$ so that $\mathbf{a} = k_1\mathbf{b} + k_2\mathbf{c}$

and **a**, **b**, **c** are coplanar.

2. Express the products in terms of determinants and the result follows.

3. As in problem 2 above.

4. Expand both sides after setting $\mathbf{a} = a_x\mathbf{i} + a_y\mathbf{j} + a_z\mathbf{k}$, $\mathbf{b} = b_x\mathbf{i} + b_y\mathbf{j} + b_z\mathbf{k}$, $\mathbf{c} = c_x\mathbf{i} + c_y\mathbf{j} + c_z\mathbf{k}$.

5. (a) 11   (b) -11 (c) 11   (d) $-\mathbf{i} + 11\mathbf{j} + 3\mathbf{k}$
(e) $13\mathbf{i} + 24\mathbf{j} - 22\mathbf{k}$ (f) $-14\mathbf{i} - 13\mathbf{j} + 25\mathbf{k}$ (g) 37
(h) $15\mathbf{i} + 8\mathbf{j} + 12\mathbf{k}$ (i) $11\mathbf{i} - 33\mathbf{j} - 11\mathbf{k}$.

6. (a) 12   (b) -12 (c) 12   (d) $-8\mathbf{i} + 8\mathbf{j} + 8\mathbf{k}$
(e) $-11\mathbf{i} + 3\mathbf{j} + 6\mathbf{k}$ (f) $3\mathbf{i} + 5\mathbf{j} + 2\mathbf{k}$ (g) 40
(h) $12\mathbf{i} - 4\mathbf{j} - 8\mathbf{k}$ (i) $24\mathbf{i} - 36\mathbf{k}$.

7. $\dfrac{\mathbf{j} \times \mathbf{k}}{\mathbf{i} \cdot \mathbf{j} \times \mathbf{k}} = \dfrac{\mathbf{i}}{1} = \mathbf{i},\ \dfrac{\mathbf{k} \times \mathbf{i}}{\mathbf{i} \cdot \mathbf{j} \times \mathbf{k}} = \dfrac{\mathbf{j}}{1} = \mathbf{j},\ \dfrac{\mathbf{i} \times \mathbf{j}}{\mathbf{i} \cdot \mathbf{j} \times \mathbf{k}} = \dfrac{\mathbf{k}}{1} = \mathbf{k}.$

8. $[\mathbf{a} \times (\mathbf{b} \times \mathbf{c}) = (\mathbf{a} \times \mathbf{b}) \times \mathbf{c}] \Leftrightarrow [(\mathbf{a} \cdot \mathbf{c})\mathbf{b} - (\mathbf{a} \cdot \mathbf{b})\mathbf{c} = (\mathbf{a} \cdot \mathbf{c})\mathbf{b} - (\mathbf{c} \cdot \mathbf{b})\mathbf{a}] \Leftrightarrow [(\mathbf{a} \cdot \mathbf{b})\mathbf{c} - (\mathbf{c} \cdot \mathbf{b})\mathbf{a} = 0] \Leftrightarrow [\mathbf{b} \times (\mathbf{c} \times \mathbf{a}) = 0].$

9. Use Identity 3.4.2 with $\mathbf{d} = \mathbf{a} \times \mathbf{b}$;

$[\mathbf{a} \cdot \mathbf{c} \times (\mathbf{a} \times \mathbf{b})]\mathbf{b} - [\mathbf{b} \cdot \mathbf{c} \times (\mathbf{a} \times \mathbf{b})]\mathbf{a}$
$$= [\mathbf{a} \cdot \mathbf{b} \times (\mathbf{a} \times \mathbf{b})]\mathbf{c} - (\mathbf{a} \cdot \mathbf{b} \times \mathbf{c})(\mathbf{a} \times \mathbf{b}),$$
$[(\mathbf{a} \times \mathbf{c}) \cdot (\mathbf{a} \times \mathbf{b})]\mathbf{b} - [(\mathbf{b} \times \mathbf{c}) \cdot (\mathbf{a} \times \mathbf{b})]\mathbf{a} = [(\mathbf{a} \times \mathbf{b}) \cdot (\mathbf{a} \times \mathbf{b})]\mathbf{c} + 0;$

now use Identity 3.4.1 to obtain the desired result.

10. By direct application of Identity 3.4.2 we have
$(\mathbf{a} \cdot \mathbf{c} \times \mathbf{v})\mathbf{b} - (\mathbf{b} \cdot \mathbf{c} \times \mathbf{v})\mathbf{a} = (\mathbf{a} \cdot \mathbf{b} \times \mathbf{v})\mathbf{c} - (\mathbf{a} \cdot \mathbf{b} \times \mathbf{c})\mathbf{v}$ and for
$\mathbf{a} \cdot \mathbf{b} \times \mathbf{c} \neq 0$, the result follows.

11. $\mathbf{a} \times [\mathbf{b} \times (\mathbf{c} \times \mathbf{d})] = \mathbf{a} \times [(\mathbf{b} \cdot \mathbf{d})\mathbf{c} - (\mathbf{b} \cdot \mathbf{c})\mathbf{d}]$
$= (\mathbf{b} \cdot \mathbf{d})(\mathbf{a} \times \mathbf{c}) - (\mathbf{b} \cdot \mathbf{c})(\mathbf{a} \times \mathbf{d}).$

12. Use the result of problem 10 with **p**, **q**, and **r** in succession. For simplicity, let
$$\mathbf{A} = \frac{\mathbf{b} \times \mathbf{c}}{\mathbf{a} \cdot \mathbf{b} \times \mathbf{c}},\ \mathbf{B} = \frac{\mathbf{c} \times \mathbf{a}}{\mathbf{a} \cdot \mathbf{b} \times \mathbf{c}},\ \mathbf{C} = \frac{\mathbf{a} \times \mathbf{b}}{\mathbf{a} \cdot \mathbf{b} \times \mathbf{c}}.$$

Now show that

$$\mathbf{p} \cdot \mathbf{q} \times \mathbf{r} = \begin{vmatrix} \mathbf{p} \cdot \mathbf{A} & \mathbf{p} \cdot \mathbf{B} & \mathbf{p} \cdot \mathbf{C} \\ \mathbf{q} \cdot \mathbf{A} & \mathbf{q} \cdot \mathbf{B} & \mathbf{q} \cdot \mathbf{C} \\ \mathbf{r} \cdot \mathbf{A} & \mathbf{r} \cdot \mathbf{B} & \mathbf{r} \cdot \mathbf{C} \end{vmatrix} (\mathbf{a} \cdot \mathbf{b} \times \mathbf{c}); \mathbf{A} \cdot \mathbf{B} \times \mathbf{C} = \frac{1}{\mathbf{a} \cdot \mathbf{b} \times \mathbf{c}}.$$

The result then follows.

13. (a) Use Identity 3.4.1.

(b) Let **a**, **b**, **c**, and **d** be unit vectors and apply the definition of vector product, using the arrangement as shown:

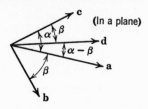

(In a plane)

14. $(\mathbf{a} \times \mathbf{b}) \times (\mathbf{c} \times \mathbf{d}) = (\mathbf{a} \cdot \mathbf{c} \times \mathbf{d})\mathbf{b} - (\mathbf{b} \cdot \mathbf{c} \times \mathbf{d})\mathbf{a} = \mathbf{0}$

since $\begin{cases} \mathbf{a} \cdot \mathbf{c} \times \mathbf{d} = 0 \Leftrightarrow \mathbf{a}, \mathbf{c}, \mathbf{d} \text{ are coplanar.} \\ \mathbf{b} \cdot \mathbf{c} \times \mathbf{d} = 0 \Leftrightarrow \mathbf{b}, \mathbf{c}, \mathbf{d} \text{ are coplanar.} \end{cases}$

The converse is false.   Let $\mathbf{a} = \mathbf{j}, \mathbf{b} = \mathbf{j}, \mathbf{c} = \mathbf{k}, \mathbf{d} = \mathbf{i}$.

15. $x = \dfrac{\mathbf{d} \cdot \mathbf{b} \times \mathbf{c}}{\mathbf{a} \cdot \mathbf{b} \times \mathbf{c}} = 3, y = \dfrac{\mathbf{d} \cdot \mathbf{a} \times \mathbf{c}}{\mathbf{b} \cdot \mathbf{a} \times \mathbf{c}} = 2, z = \dfrac{\mathbf{d} \cdot \mathbf{a} \times \mathbf{b}}{\mathbf{c} \cdot \mathbf{a} \times \mathbf{b}} = 1.$

16. $\mathbf{a} \times (\mathbf{a} \times \mathbf{b})$ is coplanar with **a** and **b** and is perpendicular to **a**. Using the result from problem 9, we have $\mathbf{b} = \frac{39}{54}\mathbf{a} - \frac{21}{54}\mathbf{u}$ where $\mathbf{u} = \mathbf{a} \times (\mathbf{a} \times \mathbf{b}) = \mathbf{i} - 4\mathbf{j} - 2\mathbf{k}$.

B. Geometric

1. Volume $= |\mathbf{a} \cdot \mathbf{b} \times \mathbf{c}| = 10.$

2. Volume $= \frac{1}{6}|\mathbf{a} \cdot \mathbf{b} \times \mathbf{c}| = \frac{5}{3}.$

3. $\overrightarrow{P_1P_2} \cdot \overrightarrow{P_1P_3} \times \overrightarrow{P_1P} = 0$ where P is a point in the plane determined by $P_1, P_2,$ and P_3.

4. $(x + 1) - 12(y + 2) - 7(z - 3) = 0$ or $x - 12y - 7z - 2 = 0.$

5. $-8x + y - 2z + 9 = 0.$

6. $55x + y - 7z - 56 = 0.$

7. $d = \dfrac{\overrightarrow{P_1P_2} \times \overrightarrow{P_1P_3}}{|\overrightarrow{P_1P_2} \times \overrightarrow{P_1P_3}|} \cdot \overrightarrow{P_1P_4}.$

8. $d = \dfrac{6}{\sqrt{29}}.$

9. $d = \frac{7}{12}\sqrt{30}.$

10. Note that $P_1(0, 0, 2)$, $P_2(3, 0, 0)$, $P_3(0, -2, 0)$ all lie in the plane $2x - 3y + 3z - 6 = 0$ so that

$$d = \frac{\overrightarrow{P_1P_2} \times \overrightarrow{P_1P_3}}{|\overrightarrow{P_1P_2} \times \overrightarrow{P_1P_3}|} \cdot \overrightarrow{P_1P} = \tfrac{29}{12} \sqrt{6}.$$

11. $d = \left| \dfrac{\overrightarrow{P_1P_2} \times \overrightarrow{P_3P_4}}{|\overrightarrow{P_1P_2} \times \overrightarrow{P_3P_4}|} \cdot \overrightarrow{P_1P_4} \right|.$

12. $d = \tfrac{2}{5} \sqrt{5}.$

13. $d = \dfrac{14}{\sqrt{202}}.$

14. $d = \tfrac{5}{3} \sqrt{3}.$

C. Physics

1. $\mathbf{N} \times \overrightarrow{P_0P_1}$ is a vector perpendicular to $\mathbf{N}$ and to $\overrightarrow{P_0P_1}$ with length equal to $|\overrightarrow{P_0P_1}|$. Hence $\mathbf{N} \times \overrightarrow{P_0P_1} \cdot \mathbf{F}$ is the projection of $\mathbf{F}$ onto $\mathbf{N} \times \overrightarrow{P_0P_1}$ time $|\overrightarrow{P_0P_1}|$, i.e., the first moment of $\mathbf{F}$ with respect to the axis AA'. Thus the vector moment is the vector $\mathbf{M} = (\mathbf{N} \times \overrightarrow{P_0P_1} \cdot \mathbf{F})\mathbf{N}$ with direction along AA' and magnitude $|\mathbf{N} \times \overrightarrow{P_0P_1} \cdot \mathbf{F}|$.

2. $\mathbf{M} = 68\mathbf{j}.$

3. $\mathbf{M} = 0\mathbf{i} = \mathbf{0}.$

4. $\mathbf{M}_1 = 5\mathbf{k}$, $\mathbf{M}_2 = 3\mathbf{k}$, and $\mathbf{M}_3 = (-2F_{3y} + F_{3x})\mathbf{k}$ to satisfy the conditions. Thus any choice of components F_{3x}, F_{3y}, F_{3z} such that $5 + 3 + (-2F_{3y} + F_{3x}) = 0$ will do, e.g., $\mathbf{F}_3 = 2\mathbf{i} + 5\mathbf{j}$.

5. $\mathbf{M}_1 = -15\mathbf{i}$, $\mathbf{M}_2 = -2\mathbf{i}$, $\mathbf{M}_3 = (7F_{3z} - 3F_{3y})\mathbf{i}$, and we must have $-15 - 2 + (7F_{3z} - 3F_{3y}) = 0$, e.g., $\mathbf{F}_3 = 6\mathbf{i} - \mathbf{j} + 2\mathbf{k}$.

6. No such force exists, since all six independent conditions cannot be satisfied by introducing a single additional force $\mathbf{F}_3$ at the point P_3.

7. See answer to problem 6.

PROBLEM SET #4: A. General

1. No solution.

2. $\mathbf{x} = \mathbf{c}.$

3. $y = \dfrac{(\mathbf{a} \times \mathbf{b}) \cdot (\mathbf{a} \times \mathbf{c})}{(\mathbf{a} \times \mathbf{b})^2}$ for $\mathbf{a}$, $\mathbf{b}$, and $\mathbf{c}$ coplanar and $\mathbf{a} \times \mathbf{b} \neq \mathbf{0}$.

4. $x = - \dfrac{(\mathbf{v} \times \mathbf{u}) \cdot (\mathbf{v} \times \mathbf{c})}{(\mathbf{v} \times \mathbf{u})^2}$ for $\mathbf{u}$, $\mathbf{v}$, and $\mathbf{c}$ coplanar and $\mathbf{v} \times \mathbf{u} \neq \mathbf{0}$.

5. Necessity: $\mathbf{a} \times \mathbf{y} = \mathbf{b} \Rightarrow \mathbf{a} \cdot \mathbf{a} \times \mathbf{y} = \mathbf{a} \cdot \mathbf{b} \Rightarrow 0 = \mathbf{a} \cdot \mathbf{b}$.

Sufficiency: (i) $\mathbf{a} \cdot \mathbf{b} = 0 \Rightarrow \mathbf{a}$ perpendicular to $\mathbf{b}$.

(ii) Choose $\mathbf{y}$ so that $\mathbf{y}$ is perpendicular to $\mathbf{b}$ and $|\mathbf{b}| = |\mathbf{a}|\,|\mathbf{y}| \sin \sphericalangle (\mathbf{a}, \mathbf{y}) \Rightarrow \mathbf{a} \times \mathbf{y} = \mathbf{b}$.

6. No, $\mathbf{y}$ can be any vector in a plane perpendicular to $\mathbf{b}$ such that $|\mathbf{b}| = |\mathbf{a}|\,|\mathbf{y}| \sin \sphericalangle (\mathbf{a}, \mathbf{y})$.

7. $\mathbf{x} = \dfrac{\mathbf{a} \times \mathbf{c} + k\mathbf{b}}{\mathbf{a} \cdot \mathbf{b}}$ for $\mathbf{a} \cdot \mathbf{b} \neq 0$.

8. $\mathbf{x} = \frac{21}{8}\mathbf{i} - \frac{15}{8}\mathbf{j} + \frac{3}{8}\mathbf{k}$.

9. $\mathbf{r} = -\frac{10}{3}\mathbf{i} - \frac{2}{3}\mathbf{j} - \frac{3}{2}\mathbf{k}$.

10. $\mathbf{r} = \frac{3}{5}\mathbf{i} - \frac{3}{10}\mathbf{j} + \frac{17}{30}\mathbf{k}$.

11. $x = 3, y = 2, z = 1$.

12. $\mathbf{x} = 3\mathbf{i} + 4\mathbf{j} + 5\mathbf{k}$.

13. $\alpha = 0, \beta = 1, \gamma = -2$ so that $\mathbf{r} = \mathbf{b} - 2\mathbf{c}$.

14. No solution, for $\mathbf{a} \cdot \mathbf{b} \times \mathbf{c} = 0$.

15. Let $\mathbf{A}, \mathbf{B}, \mathbf{C}$ be the reciprocal system to $\mathbf{a}, \mathbf{b}, \mathbf{c}$ and $\mathbf{B}', \mathbf{C}', \mathbf{D}'$ be the reciprocal system to $\mathbf{b}, \mathbf{c}, \mathbf{d}$. Then $\mathbf{r} = (\mathbf{a} \cdot \mathbf{r})\mathbf{A} + (\mathbf{b} \cdot \mathbf{r})\mathbf{B} + (\mathbf{c} \cdot \mathbf{r})\mathbf{C}$ and $\mathbf{r} = (\mathbf{b} \cdot \mathbf{r})\mathbf{B}' + (\mathbf{c} \cdot \mathbf{r})\mathbf{C}' + (\mathbf{d} \cdot \mathbf{r})\mathbf{D}'$ so that $(\mathbf{a} \cdot \mathbf{r})\mathbf{A} + (\mathbf{b} \cdot \mathbf{r})\mathbf{B} + (\mathbf{c} \cdot \mathbf{r})\mathbf{C} = (\mathbf{b} \cdot \mathbf{r})\mathbf{B}' + (\mathbf{c} \cdot \mathbf{r})\mathbf{C}' + (\mathbf{d} \cdot \mathbf{r})\mathbf{D}'$. Now take the scalar product of both sides by $\mathbf{d}$:

$$(\mathbf{a} \cdot \mathbf{r})\left(\frac{\mathbf{d} \cdot \mathbf{b} \times \mathbf{c}}{\mathbf{a} \cdot \mathbf{b} \times \mathbf{c}}\right) + (\mathbf{b} \cdot \mathbf{r})\left(\frac{\mathbf{d} \cdot \mathbf{c} \times \mathbf{a}}{\mathbf{a} \cdot \mathbf{b} \times \mathbf{c}}\right) + (\mathbf{c} \cdot \mathbf{r})\left(\frac{\mathbf{d} \cdot \mathbf{a} \times \mathbf{b}}{\mathbf{a} \cdot \mathbf{b} \times \mathbf{c}}\right) = \mathbf{d} \cdot \mathbf{r}$$

$$(\mathbf{a} \cdot \mathbf{r})(\mathbf{b} \cdot \mathbf{c} \times \mathbf{d}) + (\mathbf{b} \cdot \mathbf{r})(\mathbf{c} \cdot \mathbf{a} \times \mathbf{d}) + (\mathbf{c} \cdot \mathbf{r})(\mathbf{a} \cdot \mathbf{b} \times \mathbf{d})$$
$$= (\mathbf{d} \cdot \mathbf{r})(\mathbf{a} \cdot \mathbf{b} \times \mathbf{c})$$

and the conclusion follows.

16. (a) $\mathbf{u} = \mathbf{c} \times (\mathbf{b} \times \mathbf{a}) + (\mathbf{p} \times \mathbf{q})$ and $\alpha = 5$.

(b) $\mathbf{v} = (\mathbf{a} \times \mathbf{b}) - (\mathbf{d} \cdot \mathbf{c})\mathbf{p} - (\mathbf{c} \cdot \mathbf{p})\mathbf{d}$ and $\beta = -2$.

17. (a) $\mathbf{A} = \frac{2}{5}\mathbf{i} - \frac{1}{5}\mathbf{j} + \frac{1}{5}\mathbf{k}, \mathbf{B} = \frac{1}{5}\mathbf{i} + \frac{2}{5}\mathbf{j} + \frac{3}{5}\mathbf{k}, \mathbf{C} = \frac{1}{5}\mathbf{i} + \frac{2}{5}\mathbf{j} - \frac{2}{5}\mathbf{k}$.

(b) $\mathbf{A} = 10\mathbf{i} - 4\mathbf{k}, \mathbf{B} = 2\mathbf{i} + \mathbf{j} - \mathbf{k}, \mathbf{C} = -4\mathbf{i} + 2\mathbf{k}$.

18. Theorem 4.1.2. Use vector identities to show that

$$\mathbf{a} = \frac{\mathbf{B} \times \mathbf{C}}{\mathbf{A} \cdot \mathbf{B} \times \mathbf{C}}, \mathbf{b} = \frac{\mathbf{C} \times \mathbf{A}}{\mathbf{A} \cdot \mathbf{B} \times \mathbf{C}}, \mathbf{c} = \frac{\mathbf{A} \times \mathbf{B}}{\mathbf{A} \cdot \mathbf{B} \times \mathbf{C}}$$

Theorem 4.1.3. This theorem follows from Definition 4.1.1 and properties of the scalar product.

Theorem 4.1.4. This theorem follows from Theorem 4.1.2 and Theorem 4.1.1.

19. Theorem 4.1.5. $(\mathbf{A} \cdot \mathbf{B} \times \mathbf{C})(\mathbf{a} \cdot \mathbf{b} \times \mathbf{c})$

$$= \left(\frac{\mathbf{b} \times \mathbf{c}}{\mathbf{a} \cdot \mathbf{b} \times \mathbf{c}} \cdot \frac{\mathbf{c} \times \mathbf{a}}{\mathbf{a} \cdot \mathbf{b} \times \mathbf{c}} \times \frac{\mathbf{a} \times \mathbf{b}}{\mathbf{a} \cdot \mathbf{b} \times \mathbf{c}}\right)(\mathbf{a} \cdot \mathbf{b} \times \mathbf{c})$$

$$= \frac{1}{(\mathbf{a} \cdot \mathbf{b} \times \mathbf{c})^2}[(\mathbf{b} \times \mathbf{c}) \cdot (\mathbf{r} \times \mathbf{s})]$$

$$= \frac{1}{a \cdot b \times c^2} \begin{vmatrix} b \cdot r & b \cdot s \\ c \cdot r & c \cdot s \end{vmatrix} = \frac{(a \cdot b \times c)^2}{(a \cdot b \times c)^2} = 1$$

where $r = c \times a$ and $s = a \times b$.

Theorem 4.1.6. $A = \dfrac{j \times k}{i \cdot j \times k} = \dfrac{i}{1} = i$.   Similarly, $B = j$, $C = k$.

20. If A, B, C is the reciprocal system to a, b, c, then by Theorem 4.1.4: $r = (r \cdot a)A + (r \cdot b)B + (r \cdot c)C$ and taking the scalar product by d, the conclusion follows.

21. See answer to problem 12 in Problem Set #3.2A.

B. Geometric

1.

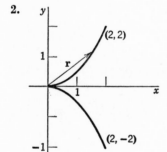

A segment of a parabola from $(0, 0)$ to $(2, 16)$ whose Cartesian form is $y = 4x^2$, $0 \le x \le 2$.

(*Note:* The scales are unequal on the graph.)

2.

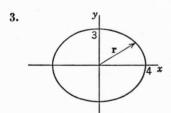

A semicubical parabola.

3.

A central ellipse with major axis 8 units and minor axis 6 units.

4.

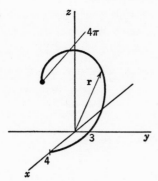

An elliptic helix.

5.

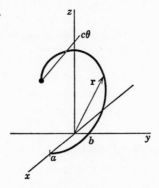

An elliptic helix.

6.

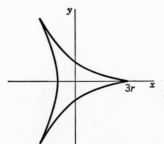

A hypocycloid of three cusps.

θ	x	y	θ	x	y
0	$3r$	0	$\dfrac{2\pi}{3}$	$-2r$	$2\sqrt{3}\,r$
$\dfrac{\pi}{2}$	$-r$	$2r$	π	$-r$	0
$-\dfrac{\pi}{2}$	$-r$	$-2r$			

7. $\mathbf{r} = 5\cos\left(\dfrac{2\pi}{3}\,t\right)\mathbf{i} + 3\sin\left(\dfrac{2\pi}{3}\,t\right)\mathbf{j}.$

8. $\mathbf{r} = 8\cos\,(8\pi t)\mathbf{i} + 5\sin\,(8\pi t)\mathbf{j}.$

9. $\mathbf{r} = \pm \dfrac{t}{\sqrt{2 - t^2}}\,\mathbf{i} + (8 - 4t^2)\mathbf{j}$ for $-\sqrt{2} < t < \sqrt{2}$

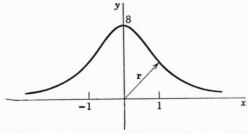

(*Note:* Scales are unequal on the graph.)

10. u is the x-coordinate of P, i.e., the directed distance of P from the yz-plane. v is the angle in the small right triangle indicated in Figure #4-1.

11. u is the z-coordinate of the point P, i.e., the directed distance of P from the xy-plane. v is the angle that the line from P to the z-axis (parallel to the xy-plane) makes with the xz-plane.

12. $u = \measuredangle\,(\mathbf{r}, \mathbf{k})$. v is the angle that the projection of $\mathbf{r}$ in the xy-plane makes with the x-axis. If $a = b = c$, the surface is a sphere.

C. Physics

1. $\mathbf{r} = (c\cos\theta + c\theta\sin\theta)\mathbf{i} + (c\sin\theta - c\theta\cos\theta)\mathbf{j}$.

2. (a) range = 800 ft, $t = 5\sqrt{2}$, $\mathbf{v} = 80\sqrt{2}\,\mathbf{i} - 80\sqrt{2}\,\mathbf{j}$

(b) maximum altitude = 200 ft, $t = \frac{5}{2}\sqrt{2}$, $\mathbf{v} = 80\sqrt{2}\,\mathbf{i}$

(c)

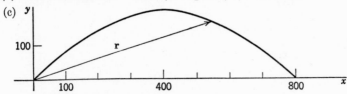

3. (a) $\mathbf{r} = 800t\mathbf{i} - 16t^2\mathbf{j}$, $\mathbf{v} = 800\mathbf{i} - 32t\mathbf{j}$

(b)

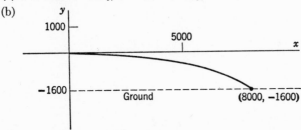

(c) $t = 10$ sec, $\mathbf{v} = 800\mathbf{i} - 3200\mathbf{j}$

4. (a) $\mathbf{r} = (400t)\mathbf{i} + (400\sqrt{3}\,t - 16t^2)\mathbf{j}$
   $\mathbf{v} = (400)\mathbf{i} + (400\sqrt{3} - 32t)\mathbf{j}$.

(b)

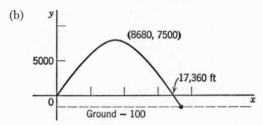

(c) t is approximately 43.5 sec
   $\mathbf{v} = 400\mathbf{i} - 699.2\mathbf{j}$

PROBLEM SET #5: A. and B. General and Geometric

1. (a) $\overrightarrow{f\left(\dfrac{\pi}{3}\right)} = \dfrac{k}{2}\mathbf{i} + \dfrac{k\sqrt{3}}{2}\mathbf{j}$ and $\overrightarrow{g\left(\dfrac{1}{12}\right)} = 5\mathbf{j} + \dfrac{\pi}{6}\mathbf{k}$ so that

$$\overrightarrow{f\left(\frac{\pi}{3}\right)} \cdot \overrightarrow{g\left(\frac{1}{12}\right)} = \frac{5\sqrt{3}}{2}\,k.$$

(b) $\overrightarrow{g\left(\dfrac{1}{18}\right)} = \dfrac{5}{2}\mathbf{i} + \dfrac{5\sqrt{3}}{2}\mathbf{j} + \dfrac{\pi}{9}\mathbf{k}$,

$\overrightarrow{F(3)} = (3v_0\cos\theta_0)\mathbf{i} + \left(3v_0\sin\theta_0 - \dfrac{9}{2}\,g\right)\mathbf{j}$

so that

$$\overrightarrow{g\left(\frac{1}{18}\right)} \times \overrightarrow{F(3)} = \left(\frac{\pi g}{2} - \frac{\pi}{3}\,v_0\sin\theta_0\right)\mathbf{i} + \left(\frac{\pi}{3}\,v_0\cos\theta_0\right)\mathbf{j}$$
$$+ \left(\frac{15}{2}\,v_0\sin\theta_0 - \frac{45}{4}\,g - \frac{15\sqrt{3}}{2}\,v_0\cos\theta_0\right)\mathbf{k}.$$

2. $\overrightarrow{g\left(\dfrac{1}{18}\right)} \cdot \overrightarrow{g\left(\dfrac{1}{6}\right)} \times \overrightarrow{g(0)} = \dfrac{5\sqrt{3}}{6}\,\pi$.

3. (a) $\overrightarrow{f(0)} = 4\mathbf{i}, \overrightarrow{f\left(\dfrac{\pi}{6}\right)} = 2\sqrt{3}\,\mathbf{i} + \mathbf{j}, \overrightarrow{f\left(\dfrac{\pi}{3}\right)} = 2\mathbf{i} + \sqrt{3}\,\mathbf{j},$

$\overrightarrow{f\left(\dfrac{\pi}{2}\right)} = 2\mathbf{k}.$

(b) and (c)

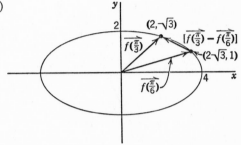

(*Note:* Vectors shown on figure are not terminated at scale points.)

4. (a) $\overrightarrow{g(0)} = \mathbf{0}, \overrightarrow{g(1)} = \mathbf{i} + \mathbf{j}, \overrightarrow{g(2)} = 2\mathbf{i} + 4\mathbf{j},$
$\overrightarrow{g(3)} = 3\mathbf{i} + 9\mathbf{j}, \overrightarrow{g(4)} = 4\mathbf{i} + 16\mathbf{j}.$

(b) and (c)

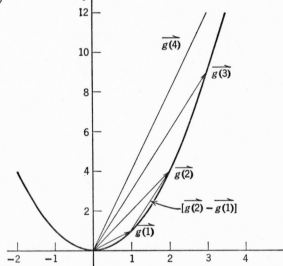

(*Note:* Unequal scales.)

5. (a) and (b)

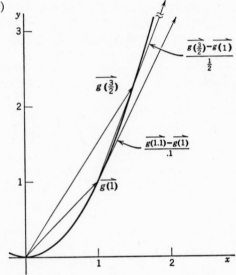

6. (a), (b), and (c)

As in problem 5 above where h "takes the successive values" such

as $\frac{1}{2}$, .1, etc. $\dfrac{\overrightarrow{g(1 + h)} - \overrightarrow{g(1)}}{h}$ approaches the vector $\mathbf{i} + 2\mathbf{j}$ as h

approaches 0 (i.e., as "h becomes as small as you please"). The
vector $\mathbf{i} + 2\mathbf{j}$ appears to be tangent to the curve at the point

$(1, 1)$, which is the terminal point of $\overrightarrow{g(1)}$.

7. (a), (b), and (c)

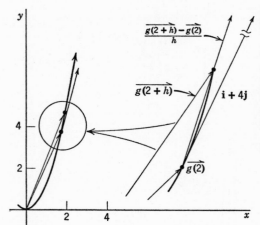

$$\frac{\overrightarrow{g(2+h)} - \overrightarrow{g(2)}}{h}$$

approaches $\mathbf{i} + 4\mathbf{j}$ as h is made small. The vector $\mathbf{i} + 4\mathbf{j}$ appears to be tangent to the curve at $(2, 4)$, the terminal end of $\overrightarrow{g(2)}$.

C. Physics

1. $p = 332{,}120$ lb/ft^2.

2.

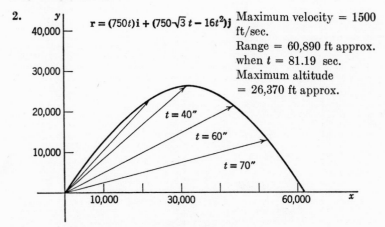

$$\mathbf{r} = (750t)\mathbf{i} + (750\sqrt{3}\,t - 16t^2)\mathbf{j}$$

Maximum velocity $= 1500$ ft/sec.
Range $= 60{,}890$ ft approx. when $t = 81.19$ sec.
Maximum altitude $= 26{,}370$ ft approx.

3. $\mathbf{F} = 1.454(10^{-10})\mathbf{r}$, $|\mathbf{F}| = 2.33$ dyne/gm.

4. $\mathbf{F} = \widehat{6}.172(10^{-12})\mathbf{r}$, $|\mathbf{F}| = 0.248$ dyne/gm.

5. $P_1(\frac{1}{4}, 4)$:
$\mathbf{v} = \frac{1}{2}\mathbf{i} - 8\mathbf{j}$,
$|\mathbf{v}| = 8.02$
$P_2(1, 1)$:
$\mathbf{v} = 2\mathbf{i} - 2\mathbf{j}$,
$|\mathbf{v}| = 2.83$
$P_3(4, \frac{1}{4})$:
$\mathbf{v} = 8\mathbf{i} - \frac{1}{2}\mathbf{j}$,
$|\mathbf{v}| = 8.02$

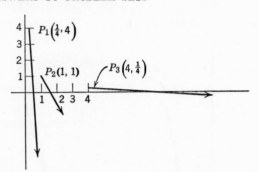

6. $P_1(\frac{1}{2}, 8)$:
$\mathbf{v} = \mathbf{i} - 16\mathbf{j}$
$|\mathbf{v}| = \sqrt{257}$
$P_2(2, 2)$:
$\mathbf{v} = 4\mathbf{i} - 4\mathbf{j}$
$|\mathbf{v}| = 4\sqrt{2}$
$P_3(8, \frac{1}{2})$:
$\mathbf{v} = 16\mathbf{i} - \mathbf{j}$
$|\mathbf{v}| = \sqrt{257}$

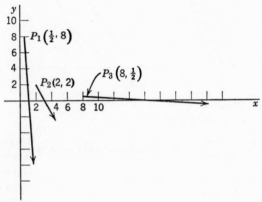

7.

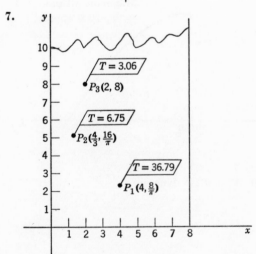

index